SpringerBriefs in Applied Sciences and Technology

Series editor

Janusz Kacprzyk, Polish Academy of Sciences, Systems Research Institute, Warsaw, Poland

W0193156

SpringerBriefs present concise summaries of cutting-edge research and practical applications across a wide spectrum of fields. Featuring compact volumes of 50 to 125 pages, the series covers a range of content from professional to academic. Typical publications can be:

- A timely report of state-of-the art methods
- An introduction to or a manual for the application of mathematical or computer techniques
- A bridge between new research results, as published in journal articles
- A snapshot of a hot or emerging topic
- An in-depth case study
- A presentation of core concepts that students must understand in order to make independent contributions

SpringerBriefs are characterized by fast, global electronic dissemination, standard publishing contracts, standardized manuscript preparation and formatting guidelines, and expedited production schedules.

On the one hand, **SpringerBriefs in Applied Sciences and Technology** are devoted to the publication of fundamentals and applications within the different classical engineering disciplines as well as in interdisciplinary fields that recently emerged between these areas. On the other hand, as the boundary separating fundamental research and applied technology is more and more dissolving, this series is particularly open to trans-disciplinary topics between fundamental science and engineering.

Indexed by EI-Compendex and Springerlink

More information about this series at http://www.springer.com/series/8884

Chunhua Sheng

Advances in Transitional Flow Modeling

Applications to Helicopter Rotors

 Springer

Chunhua Sheng
The University of Toledo
Toledo, Ohio
USA

ISSN 2191-530X   ISSN 2191-5318   (electronic)
SpringerBriefs in Applied Sciences and Technology
ISBN 978-3-319-32575-0   ISBN 978-3-319-32576-7   (eBook)
DOI 10.1007/978-3-319-32576-7

Library of Congress Control Number: 2016952517

Printed on acid-free paper

This Springer imprint is published by Springer Nature
The registered company is Springer International Publishing AG
The registered company address is: Gewerbestrasse 11, 6330 Cham, Switzerland

To my family

Haiwen, Michael and Emily

Preface

Laminar and turbulent flows are two common states of viscous fluids existing in natural environments, which have different aerodynamic and thermal characteristics. The boundary layer transition from laminar to turbulent flow is in nature a turbulence problem, one of the unsolved masteries in fluid dynamics today. Understanding the mechanism of boundary layer transition phenomena and applying it to benefit engineering designs have been great interests of scientists and engineers over the past century. With the advent of modern high performance computers as well as advanced computational modeling and simulation techniques, there has been significant progress towards an improved understanding of this fundamental fluid phenomenon. Numerical predictions of boundary layer transitions have evolved from earlier linear stability methods to more prevailing statistical modeling methods, and recently to Large Eddy Simulation (LES) or Direct Numerical Simulations (DNS).

This book provides a detailed description of numerical methods and validation processes for predicting transitional flows based on the Langtry–Menter Local Correlation-based Transition Model (LCTM), integrated with the one-equation Spalart–Allmaras (S–A) and two-equation Shear Stress Transport (SST) turbulence models. A comparative study is presented to combine the respective merits of the two coupling methods in the context of predicting the boundary layer transition phenomena from fundamental benchmark flows to realistic helicopter and tiltrotor blades. A method to correct premature flow separation is introduced in the book to address a numerical modeling issue pertinent to three-dimensional rotor aerodynamic predictions. A practical guideline is suggested for obtaining engineering solutions for realistic helicopter or tiltrotor performance using moderate computing resources.

This book will be of interest to industrial practitioners working in aerodynamic design and analysis of fixed wing or rotary wing aircraft. It will also offer advanced reading material for university graduate students in the research areas of Computational Fluid Dynamics (CFD), turbulence and transition modeling, and related fields. The structure of this book is organized as follows:

In Chap. 1, general information about the viscous fluid transition phenomena is introduced, including various transition modes and underlying mechanisms. An overview of selective predicting methods for fluid transitions is provided in Chap. 2. In Chap. 3, the Langtry–Menter's correlation-based transition model is described in detail including integration with the Spalart–Allmaras (S–A) and Menter's Shear Stress Transport (SST) turbulence models. Chapter 4 provides validations of the models in two-dimensional benchmark viscous flows, and Chap. 5 presents applications for three-dimensional realistic helicopter and tiltrotor blade performance predictions.

Toledo, USA   Chunhua Sheng

Acknowledgments

Credits go to the following individuals who have contributed or helped during the preparation of this book: my former students, Dr. Jingyu Wang, for implementing the transition model in the U^2NCLE code and generating most of the two-dimensional validation results in this book, and Dr. Qiuying Zhao, for helping process the CFD results and generating figures for this book. Special thanks go to Luke Battey for helping edit the contents and correct the language and grammar in this book. Thanks also go to Drs. Alan Wadcock and Gloria Yamauchi at NASA Ames Research Center for giving permissions to use their wind tunnel images, and Drs. Patrick Gardarein and Arnaud Le Pape at ONERA for their CFD figures used in this book. This work could not come into fruition without the support and encouragement from many people at Bell Helicopter—Dr. Matthew Hill, Dr. Albert Brand, and Mr. Tom Wood—who inspired the ideas in this book that led to the solution for JVX hover predictions. Special gratitude is extended to Jim Narramore, who retired from Bell a few years ago, for his friendship and joyful time working together on many interesting projects in the past.

Contents

Chapter 1
Introduction

Abstract Background information about boundary layer transition phenomena is described in this chapter. Various transition modes are described including natural transition, bypass transition, separation induced transition and reverse transition. Each transition mode is driven by different underlying mechanisms, and common flow parameters that influence the transition onset are discussed.

1.1 Background

Reynolds (1833) was the first to observe two different states of viscous fluid motions in his classic experiment, called laminar and turbulent flows. When the non-dimensional variable called the Reynolds number (UL/v) exceeds a certain critical value, the viscous flow starts to change its state from laminar into turbulent. This phenomenon is called viscous flow transition. When this process occurs within a boundary layer, a concept first introduced by Prandtl (1904), it is called the boundary layer transition.

For over a century, fluid dynamics researchers have devoted a vast amount of efforts in understanding this fundamental phenomenon, and have developed various theories and analytical or numerical methods to describe it. This is not only because the transition problem is in nature a turbulence problem, a century-old research topic in fluid dynamics, but also because of its prime importance in the design of advanced aircraft, ships, submarines, jet propulsions and more. The laminar and turbulent boundary layers, two common fluid states in natural environments, have rather different aerodynamic and heat transfer characteristics. Engineers have utilized the flow transition phenomenon for boundary layer control in order to improve the aerodynamic and thermal performance of the vehicle in interest. It is desired to maintain a large laminar boundary layer on a vehicle surface in order to reduce the overall drag. This is due to significantly lower skin frictions of the laminar boundary layer comparing to a turbulent boundary layer at the same Reynolds number. In addition, the effects of heat transfer are crucial for the safe operation and lifespan of reentry vehicles and gas turbine blades in jet propulsion systems.

© The Author(s) 2017
C. Sheng, *Advances in Transitional Flow Modeling*, SpringerBriefs
in Applied Sciences and Technology, DOI 10.1007/978-3-319-32576-7_1

Because the rate of heat transfer is much higher in the turbulent boundary layer than in the laminar boundary layer counterpart, it calls for a thermal management strategy using transition and flow control techniques. Contrarily, there are situations that the turbulent boundary layer is preferred, due to its better resistance to adverse pressure gradients and flow separations than the laminar boundary layer. Laminar flow is often triggered into turbulent flow prematurely in order to prevent or postpone flow separations, or improve the mixing effect and the burning efficiency of combustors in gas turbine systems.

As the boundary layer transition and flow control are highly relevant to engineering designs and applications in various industries, investigations of transition phenomena are often carried out through wind tunnel tests, theoretical analyses, computational modeling and simulations, or a combination of these methods. Sometimes flight tests may be required to verify a flow control approach or an analytical and numerical prediction. In the following sections, three common transition modes are described, followed by several common flow parameters that may influence the transition onset and process.

1.2 Transition Modes

The viscous flow transition can occur in many occasions such as boundary layer flows, shear flows and Poiseulli flows. However, the transition occurred within the boundary layer is the most complicated one and of prime importance to engineering designs such as aircraft, ships, cars, and jet propulsion systems. Because the boundary layer transition process can be triggered by different mechanisms in various applications, it is important to identify different transition modes in order to obtain basic understanding of the underlying physics.

1.2.1 Natural Transition

Natural transition, also commonly called normal transition, typically occurs in an environment with weak background noise or free-stream turbulence level (Tu). This transition is characterized by the formation of two-dimensional Tollminen-Schlichting (T-S) waves (Tollminen 1929; Schlichting 1933) in the streamwise direction, which grow in amplitude through the linear and nonlinear stages. In three-dimensional flows such as a swept wing, crossflow (C-F) waves are also developed normal to the streamwise direction, which represents an instability of inviscid type (Gregory et al. 1955). Boundary layer transition over flat plates is typically considered natural transition if the surface is perfectly smooth and the free-stream turbulence intensity is low (<1 %) (Mayle 1991).

Natural transition process in boundary layers can be divided into two stages. The first stage is the reception of disturbance waves from the free-stream or rough surface, which is denoted receptivity (Saric et al. 2002). The second stage is the growth or decay of unsteady disturbance waves inside the boundary layer. Earlier studies of natural transition were primarily based on the parallel stability theory stemming from the famous the Orr-Sommerfeld equation (OSE) (Orr 1907; Sommerfeld 1908). Herbert and Bertolotti (1987) later introduced the parabolized stability equation (PSE) in order to extend to nonparallel flows in three-dimensional compressible boundary layers. Recent studies include using Direct Numerical Simulation (DNS) to simulate the natural transition process based on supercomputers (Wu and Durbin 2001).

1.2.2 Bypass Transition

Klebanoff et al. (1962) discovered experimentally that the linear growth stage of the Tollminen-Schlichting waves can be bypassed if the magnitude of the free-stream disturbances is strong enough. This mechanism is named "bypass" by Morkovin (1984), indicating that the linear wave growth in the natural transition is irrelevant if the free-stream turbulence level is greater than 1 % (Mayle 1991). Measurements by Mayle and Schulz (1997) indicated that there is a significantly large level of unsteady velocity fluctuations in the pre-transitional flow field. The pressure fluctuation in the free-stream is believed to lead to amplification of this laminar fluctuation (Mayle and Schulz 1997).

Bypass transition is the most common transition mode in engineering applications, and is of practical importance as it signifies the departure of the skin friction from the laminar flow value. Good examples of bypass transition would be multi-stage turbine blades periodically impinged by passing wakes (Wu et al. 1999), air injection from turbine blade holes used in film cooling technologies, or transitions on helicopter rotors triggered by blade-vortex interactions (BVI). Recent DNS studies of bypass transition by Durbin and Jacobs (2002) and Brandt et al. (2004) suggest that wake-induced bypass transition is initialed by an instability in pre-transitional flow. The mean wake distortion of the boundary layer is less important to transition compared to the interaction between the boundary layer and free-stream eddies carried by the passing wakes. Of even greater interest to the engineering community are the statistical studies of free-stream disturbances in the pre-transitional boundary layer, which will be further discussed in Chap. 2.

1.2.3 Separation-Induced Transition

Separation-induced transition occurs when a laminar separation bubble forms at the leading edge of an airfoil due to adverse pressure gradients or large surface curvatures, and reattaches as turbulent flow in the downstream on the airfoil. Depending on the pressure distribution around the airfoil and other factors such as the free-stream turbulence, the laminar separation bubble can have different lengths (Mayle 1991), which have a significant impact on the lift and drag characteristics of the device. A short separation bubble on the airfoil triggers the laminar flow into turbulent due to inherently unsteady nature of laminar flows. A long separation bubble, however, can cause a thick boundary layer downstream or even lead to massive flow separation or stall on the airfoil.

Measurements of separation-induced transition are relatively easy to carry out in fixed or two-dimensional airfoils, but are more challenging in three-dimensional rotary rotors. Wadcock and Yamauchi (1998), as well as Wadock et al. (1999), observed separation-induced transition in a full-scale hovering proprotor tested at high thrusts using the oil-film interferometric skin friction technique. One challenge in the study of separation-induced transition is that it cannot be separated from both boundary layer separation and turbulence problems, for which the underlying physics is not yet well understood. Mayle and Schulz (1997) found in their experiments that the size of the separation bubble is strongly affected by the Reynolds number and the flow angle of attack. Volino and Hultgren (2001) found similar conclusions under a low-pressure gas turbine environment. Computational investigations for hovering rotors suggest that the production of turbulent eddies within the boundary layer plays a large role in determining the size of the separation bubbles (Sheng et al. 2016).

1.2.4 Reverse Transition

Reverse transition is also called relaminarization where turbulent flow is reversed back to a laminar state. This can occur when a flow acceleration parameter, $K = vU^2(dU/dx)$, reaches a level of 3.2×10^6 or higher (Mayle 1991). An airfoil flow on the leading edge of the suction surface or on the trailing edge of the pressure surface can have a strong acceleration which may cause reverse transition. An experiment of relamonarization was conducted by Savill (2002) over a flat plat with a free-stream turbulent intensity of 0.1 %, which shows the relamonarization followed by a retransition process.

1.3 Transition Parameters

It has been recognized that several flow and geometric parameters may influence the transition process. It is important to note that a transition process is often triggered by combined effects of different factors, whose influence can vary in different situations. Therefore, it would be beneficial to understand how these parameters influence the transition onset and process. This would not only help scientists develop enhanced theories and models to describe the transition phenomena, but also engineers designing optimal flow control strategies in practical engineering applications.

1.3.1 Free-Stream Turbulence

The free-stream turbulence level (Tu) is known to have a large influence on all transition modes. A high level of the free-stream turbulence often triggers bypass transitions, which are the most common transition modes in engineering applications. It also influences separation-induced transitions, and sometimes determines whether or not a laminar separation bubble will reattach as turbulent flow or burst into massive separation. Numerical investigations indicated that the influence of free-stream turbulence on the transition onset is weakened under strong pressure gradients, as demonstrated in the numerical computations for highly twisted proprotors by Sheng and Zhao (2016).

1.3.2 Pressure Gradient

Along with the free-stream turbulence, the pressure gradient is another important parameter that can strongly influence the transition onset and process, or even the flow separation. In general, a favorable pressure gradient serves to defer the transition onset while an adverse pressure gradient promotes the onset of transition. A reverse transition phenomenon can occur under a very strong favorable pressure gradient (Mayle 1991). While the effect of pressure gradients on the fluid transition and separation is well understood, the combined effects of pressure gradients with other parameters, such as the turbulence level outside boundary layers or the turbulent eddy within the boundary layer, may complicate the transition process.

1.3.3 Surface Roughness

When a surface roughness is small (smooth wall) and the background noise is weak, natural transition occurs in most scenarios. When the surface roughness is large enough, the bypass transition may be triggered. Boyle and Senyitko (2003) investigated the surface roughness effects on the loss and transition of turbine vane aerodynamics, and concluded that the roughness effects strongly depend on Reynolds numbers. Roberts and Yaras (2005) concluded that the effect of surface roughness is comparable to that of the free-stream turbulence, and it also becomes an important factor in determining the separation-induced transition. Helicopter or tiltrotor blade tips are often treated with rough structural materials in order to trip the laminar boundary layer into turbulent flow.

1.3.4 Unsteady Velocity Fluctuation

It has been recognized that a high level of fluctuation in streamwise velocity profiles is the cause of wake-induced bypass transition and breakdown into a fully turbulent flow in a pre-transitional laminar flow (Wu and Durbin 2001). It should be noted that the streamwise fluctuation is different from the turbulent fluctuation, such that large eddies near the wall contribute to the production of non-turbulent fluctuations while small eddies within the boundary layer contribute to the turbulence production (Mayle and Schultz 1997).

1.3.5 Turbulent Viscosity

Free-stream turbulent viscosity is often represented as a ratio of the turbulent eddy viscosity to the molecular viscosity. Its influence on viscous flow transition is reflected as a decay rate of turbulence in the flow field. A recent numerical study by Spalart and Rumsey (2007) indicated that a lower freestream eddy viscosity level causes a more rapid decay of turbulence in the flow field. In addition, the decay rates of the turbulent kinetic energy and the turbulent frequency are much higher than that of the eddy viscosity, which in turn affect the transition onset and process.

1.3.6 Other Factors

The effect of compressibility on transitions is weak (Boyle and Simon 1999). However, convective mass and heat transfers, such as used in film cooling and heat exchangers, affect the transition Reynolds number and thus the transition process. Mass flow injection serves to trigger the bypass transition, while mass extraction may defer or trigger the transition onset as well. Studies on mass and heat transfers and their influences on transitions (including reverse transition) are of practical importance in the design of flow control technologies for engineering applications.

References

Boyle RJ, Senyitko RG (2003) Measurement and prediction of surface roughness effects on turbine vane aerodynamics. In: Proceedings of ASME TURBO EXPO 2003, GT-2003–38580, Atlanta, Georgia, 16–19 June 2003

Boyle RJ, Simon FF (1999) Mach number effects on turbine blade transition length prediction. J Turbomach 121(4):694–702. doi:10.1115/1.2836722

Brandt L, Schlatter P, Henningson DS (2004) Transition in boundary layers subject to free-stream turbulence. J Fluid Mech 517:167–198

Durbin PA, Jacobs RG (2002) DNS of bypass transition. In: Closure strategies for turbulent and transitional flow

Gregory H, Stuart JT, Walker WS (1955) On the stability of three-dimensional boundary layers with applications to the flow due to a rotating disk. Philos Trans R Soc London Ser A 248:155–199

Herbert T, Bertololli FP (1987) Stability analysis of nonparallel boundary layers. Bull Am Phys Spc 32:2079–2806

Klebanoff PS, Tidstrom KD, Sargent LM (1962) The three-dimensional nature of boundary layer instability. J Fluid Mech 12:1–34

Mayle RE (1991) The role of laminar turbulent transition in gas turbine engines. Trans ASME: J Turbomachinery 113:509–537

Mayle R, Schultz A (1997) The path to predicting bypass transition. J Turbomach 119:405–411

Morkovin MV (1984) Bypass transition to turbulence and research desiderata. In: Graham RW (ed) Transition in turbines, NASA CP-2386, pp 161–204, 1884

Orr W (1907) The stability or instability of the steady motions of a perfect liquid and of a viscous fluid. In: Proceedings of the Royal Irish Academy. Section A, p 27

Prandtl L (1904) Uber flussigkeitsbewegung bei sehr kleiner reibung. Verhandlungen III Intern Math Kongress, Heidelberg, 1904, S. 484 (Crouch JD, Herbert T (1986) Perturbation analysis of nonlinear secondary instability in boundary layers. Bull Am Phys Soc 31:1718)

Reynolds O (1833) An experimental investigation of the circumstances which determine whether the motion of water should be direct or sinuous, and the law of resistance in parallel channels. Philos Trans R Soc London 174:935–982

Roberts SK, Yaras MI (2005) Boundary-layer transition affected by surface roughness and free-stream turbulence. J Turbomach 127:449–457

Saric WS, Reed HL, Kerschen EJ (2002) Boundary-layer receptivity to freestream disturbances. Annu Rev Fluid Mech 34:291–319

Savill AM (2002) New strategies in modeling by-pass transition. In: Launder BE, Sandsam ND (eds) Closure strategies for turbulent and transitional flows. Cambridge University Press, Cambridge, pp 464–492

Schlichting H (1933) Zur entstehung der turblenz bei der plattenstromung. Nachr. Ges Wiss. Gottingen, Math-phys, KL

Sheng C, Zhao Q (2016) Assessment of transition models in predicting the skin frictions and flow field of a full-scale tilt rotor in hover. In: Proceedings of the AHS 72nd annual forum, West Palm Beach, Florida, 16–19 May 2016

Sheng C, Wang J, Zhao Q (2016) Improved rotor hover predictions using advanced turbulence modeling. J Aircraft. 53(5):1549–1560. doi:http://dx.doi.org/10.2514/1.C033512

Sommerfeld A (1908) Ein beitrag zur hydrodynamischen erklaerung der turbulenten fluessigkeits-bewegungen. In: Proceedings of 4th international congress of mathematics, vol III, Rome

Spalart PR, Rumsey CL (2007) Effective inflow conditions for turbulence models in aerodynamic calculations. AIAA J 45(10):2533–2544

Tollmien W (1929) Uber die entstehung der turbulenz. Nachr. Ges Wiss. Gottingen, Math-phys, KL

Volino RJ, Hultgren LS (2001) Measurements in separated and transitional boundary layers under lo-pressure turbine airfoil conditions. J Turbomach 123:189–197

Wadcock AJ, Yamauchi GK (1998) Skin friction measurements on a full-scale tilt rotor in hover. In: Proceedings of AHS 54th annual forum, Washington, DC, 20–22 May 1998

Wadcock AJ, Yamauchi GK, Driver DM (1999) Skin friction measurements on a full-scale tilt rotor in hover. J Am Helicopter Soc 44(4):312–319

Wu X, Durbin PA (2001) Evidence of longitudinal vortices evolved from distorted wakes in a turbine passage. J Fluid Mech 446:199–228

Wu X, Jacobs RG, Hunt JCR et al (1999) Simulation of boundary layer transition induced by periodically passing wakes. J Fluid Mech 398:109–153

Chapter 2
Transition Prediction

Abstract A selective review of methods for transition modelling and simulation is provided in this chapter. Methods are grouped into three major categories: analytical models based on the stability theory, transition models based on statistical Reynolds-averaged Navier-Stokes (RANS) equations, and Large Eddy Simulation (LES) and Direct Numerical Simulation (DNS). The advantages and disadvantages of the methods in each category are assessed for their compatibility for use in general purpose engineering CFD codes.

2.1 Overview

Transition is a complex phenomenon that involves multiple mechanisms in different applications. The nature of the transition problem is also the origin of the turbulence problem, one of the unsolved mysteries in fluid dynamics today. Because of its practical significance in engineering designs such as airplanes, ships, cars, and spacecraft, etc., significant efforts have been devoted to transition and turbulence research over the past decades. With the recent advent of high performance computing architectures and numerical solution algorithms, computational modelling and simulation has increasingly become an important tool both in scientific research and engineering designs in these fields. Efforts of different research groups have resulted in a spectrum of transition modelling approaches that can be used in different applications with acceptable accuracy. However, challenges still exist in developing robust and reliable transition models for practical engineering applications. For example, the transition process involves a wide range of scales in time and length, as well as linear and nonlinear interactions between free-stream properties and boundary layer dynamics. However, these coherent interactions may be flitted out through the Reynolds averaging of the Navier-Stokes equations in most practical CFD codes (Stock and Haase 2000). In addition to numerical limitations, it is virtually impossible to include all mechanisms into a physics-based equation framework (Menter et al.

© The Author(s) 2017
C. Sheng, *Advances in Transitional Flow Modeling*, SpringerBriefs
in Applied Sciences and Technology, DOI 10.1007/978-3-319-32576-7_2

2015), due to a lack of complete understanding of how the underlying physics drives the transition process in various applications. Furthermore, the transition modelling cannot be isolated from the turbulence problem and flow separation, which often results in coherent interferences among different models (Savill 1993a, b).

Due to the complexity of transition phenomena, methods to predict transition have been evolved into three major categories over the past decades. The first category includes methods based on stability theory to predict the natural transition, stemmed from the Prandtl's small disturbance hypothesis (Prandtl 1904). The most noticeable one in this group is the e^N method, developed by Smith and Gamberoni (1956). Methods in the second category are based on statistical modelling of RANS equations to predict bypass and separation-induced transitions, which comprise the majority of transition modeling research in the CFD community today. The central concepts of statistical modelling for transition are stemming from Emmons's classic work (1951) of turbulent spots as well as the intermittency distributions of Dhawan and Narasimha (1958). Examples of this class include the low Reynolds number turbulence closure methods (Wilcox 1992), the intermittency method with experimental correlations (Suzen and Huang 2000, Menter et al. 2002), the laminar kinetic energy method (Walters and Leylek 2002, 2004), and many others. Methods in the last category include direct numerical simulation (DNS) (Durbin and Jacobs 2002) and large eddy simulation (LES) (Wu et al. 1999) to simulate various transition processes directly in order to avoid or minimize the usage of turbulence closure models.

The motivation of this book is to document recent advances in statistical transition modelling based on state-of-the-art parallel CFD methods that are suitable for practical engineering applications. A major method described in this book is the Local Correlation-based Transition Model, or $\gamma - \widetilde{Re}_{\theta t}$ model, developed by Menter et al. (2002, 2004, 2006) and Langtry and Menter (2006, 2009) over the last decade. Although a simplified model was recently proposed by Menter et al. (2015) to solve only the γ—equation, it is not within the scope of this book. The $\gamma - \widetilde{Re}_{\theta t}$ model (Langtry and Menter 2006, 2009) solves two transport equations: one for intermittency (γ) and another for transition momentum-thickness Reynolds number ($\widetilde{Re}_{\theta t}$). It was originally constructed under the framework of Menter's two-equation Shear Stress Transport (SST) turbulence model (Menter 1994), and was later integrated into the one-equation Spalart-Allmaras (S-A) turbulence model (Spalart and Allmaras 1994) by Medida and Baeder (2011, 2013a, b) and by Wang and Sheng (2014, 2015), respectively. A new method to correct premature flow separations, a common issue encountered in rotor CFD simulations (Sheng 2014; Sheng et al. 2016), is detailed in this book to order to improve the models' behaviour near separation points. The purpose of this book is to provide complete numerical procedures for implementing the Langtry and Menter $\gamma - \widetilde{Re}_{\theta t}$ transition model into a modern high-resolution implicit numerical scheme, and to offer advanced reading materials for university graduate students working in research areas of CFD, turbulence, and transition modeling. It also provides a practical guideline for industrial practitioners in aerodynamic design and analysis for fixed-wing or rotary wing aircraft using modern CFD solvers based on moderate computer resources.

Following this chapter, an overview is provided for selected transition modeling methods developed in the aforementioned three categories. There are some good review articles and papers published in this subject such as Aupoix et al. (2011), Pasquale et al. (2009), Sevningsson (2006) and more. While the current work should not be considered all-inclusive for the efforts of transition modeling devoted over the past decades, it is the authors' hope that this collection of relevant work will serve as a starting point for further research in this fundamental and important field of fluid dynamics.

2.2 Methods Based on Stability Theory

Methods in this category are stemming from Prandtl's small disturbance hypothesis (Prandtl 1904), with the intention of predicting the natural transition onset occurred in low free-stream turbulence environments. Two representative methods are the e^N method developed by Smith and Gamberoni (1956) and van Ingen (1956), and the parabolized stability equations (PSE) method of Herbert and Bertolotti (1987).

2.2.1 The e^N Method

The e^N method, developed by Smith and Gamberoni (1956) and van Ingen (1956) over 50 years ago, is one of the most popular methods in this category. This method is based on the linear stability theory with a local parallel flow assumption, and calculates the growth of disturbance amplitudes from the boundary layer neutral point to the transition location. The factor N is the total growth rate of the most unstable disturbances. The application of the e^N method involves three steps: the first step is to calculate the laminar velocity and temperature profiles at different stream-wise locations for the given geometry of interest. The second step is to calculate the local growth rates of the most unstable waves among all velocity profiles. This leaves the last step being to calculate the transition onset location based on the local growth rates integrated along each streamline.

The major problem with the e^N method is that due to its local parallel flow assumption it cannot predict the transition caused by nonlinear effects such as bypass or surface roughness-induced transitions. In addition, the value of N for the transition onset is not universal and needs to be correlated with the experimental data (Warren and Hassan 1997). Hence, the e^N method is considered as a semi-empirical method at best. Finally, the e^N method is not compatible with the most engineering CFD methods in use today, because it requires solving the boundary layer equations for global flow quantities. Arthur and Atkin (2006) developed a procedure by applying the e^N method within a conventional RANS framework. Firstly, an initial guess of the transition onset location was obtained based on the e^N criterion and a series of pressure distributions were extracted from the RANS solution. The improved boundary layer profiles were then fed back to the stability analysis to yield a new transition onset location, until the RANS solution and transition location are fully converged. Aupoix et al. (2011) implemented simplified stability methods called "database methods" at ONERA. However, these stability-based methods were difficult to apply on massively parallel computers, due to requirements of non-local quantities such as the boundary layer momentum thickness or shape factor (Aupoix et al. 2011).

2.2.2 Parabolized Stability Equation Method

Driven by the need to consider the nonparallel effects neglected in the linear stability theory, Herbert and Bertolotti (1987) proposed parabolized stability equations (PSE) methods. The mean flow, amplitude functions, and wave numbers were calculated based on the streamwise distance to predict the transition onset location. Savill (2002) developed the nonlinear parabolized stability equations in order to predict the subsequent transition region after the transition onset point. At first sight, these linear and nonlinear PSE methods could be considered to solve both linear and nonlinear development of disturbance waves. However, the growth of disturbance amplitude is required to evaluate along the streamlines, which is a significant difficulty for three-dimensional flow computations because the streamline direction is not always aligned with the body surface. Furthermore, development of disturbance waves highly depends on the initial amplitude of the waves, which is not universal just like the N-factor in the e^N method. These limitations make the method difficult for predicting the transition onset in three-dimensional realistic flows with complex geometries in engineering applications.

2.3 Statistical Methods of Transition Modelling

All statistical modeling methods are stemming from the classic work of Emmons (1951) on the formation of turbulent spots and the intermittency distributions of Dhawan and Narasimha (1958). Differentiations in these methods exist in the way of obtaining the intermittency distributions. This is either done by using algebraic correlations with experimental data, solving transport equations, or could also be a combination of both. Other modeling methods include the low Reynolds number turbulence models (Wilcox 1992) and the laminar kinetic energy methods (Walters and Leylek 2002, 2004).

2.3.1   Low Reynolds Number Turbulence Models

The concept of using low Reynolds number turbulence models to predict transition is based on the models' wall damping capability in the boundary layer (Jones and Launder 1973) to simulate the transition process, such as bypass transition that is dominated by a diffusion effect from the freestream. These models typically suffer from a close interaction between the transition capability and the viscous sublayer modeling, which prevents independent calibration of both phenomena (Savill 1993a, b). There are several low Reynolds number models where transition prediction was specifically considered during the model calibration, such as the Wilcox low Reynolds $k - \omega$ model (Wilcox 1992), the Langtry and Sjolander's low Reynolds number $k - \varepsilon$ model (Langtry and Sjolander 2002), and Walters and Leylek's transition model (Walters and Leylek 2002, 2004). Regardless, these models still exhibit a close connection between the sublayer behavior and the transition calibration. Re-calibration of one model would change the performance of the other (Menter and Langtry 2006). Models like Launder and Sharma model (1974), where the near-wall behavior is described by the turbulent Reynolds number, perform better than those that use the local wall distance. However, no model in this group provides reliable transition predictions for any combination of Reynolds numbers, free-stream turbulence levels, and pressure gradients (Pasquale et al. 2009).

2.3.2   Correlation-Based Intermittency Models

Models in this group use the concept of intermittency (γ) of Dhawan and Narasimha (1958) to blend the laminar and turbulent flow regimes. Intermittency γ is a measure of the probability that a given point in space is located inside the turbulent region. In other words, it is the fraction of time that the flow is turbulent during transition. By setting the intermittency factor as ranging from zero to one, the value of zero then represents the pre-transitional laminar flow and the value of one being the fully turbulent flow. In practice, this intermittency factor is multiplied to the production source term of turbulent eddy viscosity, which is calculated by Reynolds-averaged Navier-Stokes equations (RANS) codes. This approach neglects the interaction between the non-turbulent and turbulent parts of the flow during the transition, but the loss of some physical information is acceptable in the content of statistical modeling using the RANS codes (Pasquale et al. 2009).

Two selections need making in order to utilize the intermittency-based approaches for transition prediction. One is how to determine the intermittency factor distributions, and the second is how to define the transition onset criterion. Various methodologies have been proposed over the past decades, which form the main body of transition modeling methods in this category. The intermittency factor can generally be determined by algebraic models (Gostelow et al. 1994) or by transport equation (Suzen and Huang 2000; Suzen et al. 2002; Menter et al. 2002). The transition onset criterion can be determined based on empirical or experimental correlations (Abu-Ghannam and Shaw 1980) or based on the transport equation with correlation (Langtry 2006; Langtry and Menter 2009).

The algebraic intermittency models were widely used in earlier structured grid CFD codes (Arnel 1988) where the transition onset location was usually based on the empirical correlation of Abu-Ghannam and Shaw (1980). Suzen and Huang (2000) proposed a new approach based on the transport equation to solve the intermittency factor, where the source terms were designed to mimic the behavior of some algebraic intermittency models. Similar efforts were made by Menter et al. (2002) and Steeland and Dick (2001). The transition onset criterion was determined by the local Reynolds number and the transition onset Reynolds number. Both numbers were functions of the free-stream turbulence intensity and the acceleration parameter. These non-local quantities were compared with the experimental correlation of Huang and Xiong (1998) to determine the transition onset location. These models were validated for flows with zero-pressure and adverse-pressure gradients at different free-stream turbulence intensities, and received good agreements with the experimental data of Savill (1993a, b). The major problem of this transition model is the requirement of non-local quantities, which makes it difficult to be applied for three-dimensional flows or using unstructured grid parallel CFD solvers.

The transition model, proposed by Papp and Dash (2005), was based on a concept analogous to the Warren, Harris and Hassan one-equation model (Warren et al. 1995). An additional transport equation was solved for the non-turbulent fluctuations that include the cross-flow instabilities and second mode instabilities. The transition onset location was predicted as the minimum distance along the surface. This distance was determined by the turbulent viscosity coefficient, kinematic viscosity and eddy viscosity due to the non-turbulent fluctuations. The Papp and Dash transition model was implemented into the RANS solver by multiplying the turbulent eddy viscosity with the intermittency. Simulations showed that the transition onset location was properly obtained, but the peak value in heat transfer did not match correctly in some cases. This discrepancy was attributed to the algebraic nature of the intermittency function use.

Probably the most notable model in this category is the so-called Local Correlation-based Transition Model, or $\gamma - \widetilde{Re}_{\theta t}$ model, proposed by Menter and Langtry about a decade ago (Menter et al. 2002, 2004, 2006; Langtry and Menter 2009). This model solves two transport equations: one for intermittency (γ) and another for the momentum thickness Reynolds number ($\widetilde{Re}_{\theta t}$). In this formulation, only local information is used to activate the production term in the intermittency equation. The link between the intermittency equation and the correlation is achieved through the use of vorticity Reynolds number ($Re_{\theta t}$), which only depends on local variables such as density, viscosity, vorticity, and local wall distance.

Initially, only the conceptual framework of this model was published by Menter et al. (2002, 2004, 2006). A full disclosure of the $\gamma - \widetilde{Re}_{\theta t}$ model was later published by Langtry and Menter (2009), including all experimental correlations used to determine the transition onset location for different boundary layer velocity profiles. The $\gamma - \widetilde{Re}_{\theta t}$ model, however, is not considered to satisfy the Galilean invariance (Menter et al. 2015), which limits the model from being applied to surfaces moving relative to the coordinate system. Menter et al. (2015) recently proposed a simplified one-equation γ model in order to overcome the deficiency in the original $\gamma - \widetilde{Re}_{\theta t}$ model. While the $\gamma - \widetilde{Re}_{\theta t}$ transition model was originally solved under the framework of the Menter two-equation SST turbulence model (Menter 1994), extensions were made by incorporating it into a widely used one-equation Spalart-Allmaras (S-A) turbulence model by Medida and Baeder (2011, 2013a, b) under a structured grid RANS framework, and by Wang and Sheng (2014) under an unstructured grid RANS framework. In addition, extensions to include crossflow instability in the $\gamma - \widetilde{Re}_{\theta t}$ model were recently proposed by several researchers such as Seyfert and Krumbein (2012), Medida and Baeder (2013a, b), Grabe and Krumbein (2014), and Grabe et al. (2016). All these extensities show the versatility of the $\gamma - \widetilde{Re}_{\theta t}$ model in any correlation based formulations.

2.3.3   The Laminar Kinetic Energy Method

In contrast to the models using the intermittency factor, a different approach was proposed by Walters and Leylek (2002) that solves the transport equation for laminar kinetic energy. This method is based on the concept that bypass transition is caused by very high amplitude streamwise fluctuations, which are very different from turbulent fluctuations. Mayle and Schulz (1997) proposed a second kinetic energy equation to describe these streamwise fluctuations, called laminar kinetic energy (K_L). In the near wall region, the turbulent kinetic energy (K_T) is split into small-scale energy and large-scale energy. The small-scale energy contributes directly to the turbulence production and the large-scale energy contributes to the production of non-turbulent fluctuations (K_L). Volino (1998) believed that the amplification of laminar kinetic energy (K_L) is caused by redirection of normal velocity fluctuation into streamwise direction, which generates local pressure gradients in the boundary layer and leads to breakdown of laminar fluctuations into full turbulent flows. The transition onset in the Walters and Leylek model was determined by a parameter that is based on the turbulent kinetic energy, the kinetic eddy viscosity, and the wall distance. The validations of the model were performed based on low Reynolds number $\kappa - \varepsilon$ model, which yielded a reasonable prediction of the transition onset. This is also a single-point transition model that only needs local flow quantities. The major issue of this model is that the calibration of the transition model will affect the solution in fully turbulent flows, and the model is not flexible enough for a wide range of transition mechanisms in realistic applications.

Similarly, Lodefier et al. (2006) also proposed a model based on the concept of streamwise fluctuations, but introduced the intermittency equation to describe the transition region. The intermittency equation was based on the work of Steelant and Dick (2001), which was multiplied to the production term to start the transition process. Like Langtry and Menter's model, the vorticity Reynolds number was used to trigger the transition. However, the model used the free-stream turbulence intensity to evaluate the transition-momentum thickness Reynolds number, and the empirical correlation used for the transition momentum thickness Reynolds number does not include a pressure gradient. This model was incorporated into the SST $k - \omega$ model by multiplying the eddy viscosity with the intermittency and the modification of production terms in the turbulent kinetic energy and dissipation rate equations. Unlike the Langtry and Menter's model, this model used the free-stream turbulence intensity to determine the transition onset and was not a single point model.

2.4 Transition Simulation Methods

Direct Numerical Simulation (DNS) solves the fully unsteady Navier-Stokes equations directly, and does not need any turbulence models to close the equations. Theoretically, it can simulate the whole transition process including the development of disturbance waves, interaction between waves and boundary layers, ignition of turbulent spots, laminar flow breakdown, and development into fully turbulent flow (Durbin and Jacobs 2002). However, the grid for the DNS computations has to be extremely fine in order to capture the small scales of turbulent flows. The total number of grid points is in the scale of $O(Re_L^3)$ for a "modelling free" simulation (Aupoix et al. 2011). With the increasing speed of modern CPU's and the advent of cluster computing, DNS computations have moved beyond simple flat plates and it is now possible to perform DNS computations of a three-dimensional low-pressure turbine blade at Reynolds numbers up to 1.5×10^5 (Wu and Durbin 2001). However, from a computing resource standpoint it is still prohibitive to use DNS to simulate practical engineering problems, such as full vehicle configurations operating at high Reynolds numbers.

Due to significant computational costs associated with DNS, Large Eddy Simulation (LES) (Wu et al. 1999) is an alternative method for many researchers who have tried to solve transitional flows. LES uses the concept of solving the large-scale eddies directly, and modeling the small eddies using the Smagorinsky eddy viscosity approach (Smagorinsky 1963). This may create a major problem, as the transition onset location predicted by LES is sensitive to the value of Smagorinsky constant. This constant is needed to calibrate the local sub-grid eddy viscosity. The dynamic sub-grid-scale model developed by Germano (1992) was more appropriate for predicting the transition onset, because the sub-grid eddy viscosity was automatically reduced to zero in a laminar boundary layer.

Because DNS methods solve all scale levels of turbulence to the smallest grid sizes, they are capable of simulating various transition processes including natural transition, bypass transition, and separation-induced transitions, etc. However, the computational costs for DNS or LES methods are prohibitively high in solving problems for realistic geometries at high Reynolds numbers. Therefore, these methods for transition simulations are largely used as research tools in academies or as substitutes for controlled experiments.

References

Abu-Ghannam BJ, Shaw R (1980) Natural transition of boundary layers—the effects of turbulence, pressure gradient, and flow history. J Mech Eng Sci 22(5):213–228

Arnel D (1988) Laminar-turbulent transition problems in supersonic and hypersonic flows. In: AGARD-FDP-VKI special course on aerothermodynamics of hypersonic vehicles

Arthur MT, Atkin CJ (2006) Transition modeling for viscous flow prediction. Paper presented at 36th AIAA fluid dynamics conference, AIAA paper 2006-3052, San Francisco, California

Aupoix B, Arnal D, Bezard H et al (2011) Transition and turbulence modeling. J AerospaceLab (2)

Dhawan D, Narasimha R (1958) Some properties of boundary layer flow during transition from laminar to turbulent motion. J Fluid Mech 3:418–436

Durbin PA, Jacobs RG (2002) DNS of bypass transition. In: Launder BE, Sandsam ND (eds) Closure strategies for turbulent and transitional flow. Cambridge University Press, Cambridge, pp 464–492

Emmons H (1951) The laminar-turbulent transition in a boundary layer—part I. J Aeronaut Sci 18:490–498

Germano M (1992) Turbulence: the filtering approach. J Fluid Mech 238:325–336

Gostelow JP, Blunden AR, Walker GJ (1994) Effects of free-stream turbulence and adverse pressure gradients on boundary layer transition. ASME J Turbomach 116:392–404

Grabe C, Krumbein A (2014) Extension of the $\gamma - Re_{\theta t}$ model for prediction of crossflow transition. AIAA paper 2014-1269. Paper presented at 52nd aerospace sciences meeting, National Harbor, Maryland

Grabe C, Shengyang N, Krumbein A (2016) Transition transport modeling for the prediction of crossflow transition. AIAA paper 2016-3164. Paper presented at 34th applied aerodynamics conference, Washington DC, June 2016

Herbert T, Bertololli FP (1987) Stability analysis of nonparallel boundary layers. Bull Am Phys Spc 32:2079–2806

Huang PG, Xiong G (1998) Transition and turbulence modeling of low pressure turbine flows. AIAA 98-0339. Paper presented at the 26th AIAA aerospace sciences meeting, Reno, Nevada

Jones WP, Launder BE (1973) The calculation of low Reynolds number phenomena with a two-equation model of turbulence. Int J Heat Mass Transfer 15:301–314

Langtry RB (2006) A correlation-based transition model using local variables for unstructured parallelized CFD codes. PhD Dissertation, University of Stuttgart

Langtry RB, Menter FR (2006) Predicting 2D airfoil and 3D wind turbine rotor performance using a transition model for general CFD codes. Paper presented at 44th AIAA aerospace sciences meeting and exhibit, Reno, Nevada

Langtry RB, Menter FR (2009) Correlation-based transition modeling for unstructured parallelized computational fluid dynamics codes. AIAA J 47(12):2894–2906

Langtry RB, Sjolander SA (2002) Prediction of transition for attached and separated shear layers in turbomachinery, AIAA Paper 2002-3643. Paper presented at 38th AIAA/ASME/SAE/ASEE joint propulsion conference, Indianapolis, Indiana

Launder BE, Sharma BI (1974) Application of the energy-dissipation model of turbulence to the calculation of flow near a spinning disc. Lett Heat Mass Transfer 1(2):131–138

Lodefier K, Merci B, De Langhe C et al (2006) Intermittency based RANS transition modeling. Progress Comput Fluid Dyn 6(1/2/3)

Mayle R, Schultz A (1997) The path to predicting bypass transition. J. Turbomach 119:405–411

Medida S, Baeder JD (2011) Application of the correlation-based $\gamma - \overline{Re_{\theta t}}$ transition model to the Spalart-Allmaras turbulence model. In: 20th AIAA computational fluid dynamics conference, Honolulu, Hawaii, June 2011

Medida S, Baeder JD (2013a) Role of improved turbulence and transition modeling methods in rotorcraft simulations. In: Proceedings of the AHS international 69th annual forum, Phoenix, Arizona, May 2013

Medida S, Baeder JD (2013b) A new crossflow transition onset criterion for RANS turbulence models. AIAA paper 2013-3081. Paper presented at the 21st AIAA Computational Fluid Dynamics Conference, San Diego, California 2013

Menter FR (1994) Two-equation eddy-viscosity turbulence models for engineering application. AIAA J 32(8):1598–1605

Menter FR, Esch T, Kubacki S (2002) Transition model based on local variables. Paper presented at 5th international symposium on engineering turbulence modelling and measurement, Mallorca, Spain

Menter FR, Langtry RB, Likki SR (2004) A correlation based transition model using local variables part I—model formulation. ASME Paper No. GT-2004-53452

Menter FR, Langtry RB, Volker S (2006) Transition modelling for general purpose CFD codes. J. Flow Turbul Combust 77:277–303

Menter FR, Smirnov PE, Liu T et al (2015) A one-equation local correlation-based transition model. Flow Turbul Combust. doi:10.1007/s10494-015-9622-4

Papp JL, Dash SM (2005) Rapid engineering approach to modeling hypersonic laminar-to-turbulent transitional flows. J Spacecraft Rockets 42(3):467–475

Pasquale DD, Rona A, Garrett SJ (2009) A selective review of CFD transition models. Paper presented at 39th AIAA Fluid Dynamics Conference, San Antonio, Texas, 22–25 June 2009

Prandtl L (1904) Uber flussigkeitsbewegung bei sehr kleiner reibung. Verhandlungen III Intern Math Kongress, Heidelberg, 1904, S. 484 (Crouch J D and Herbert T (1986) Perturbation analysis of nonlinear secondary instability in boundary layers, Bull Am Phys Soc, 31:1718)

Savill AM (1993a) Some recent progress in the turbulence modeling of by-pass transition. In: So RMC, Speziale CG, Launder BE (eds) Near-wall turbulent flows. Elsevier, pp 829–848

Savill AM (1993b) Evaluating turbulence model predictions of transition. Advances in Turbulence IV. Springer, Netherlands, pp 555–562

Savill AM (2002) New strategies in modeling by-pass transition. In: Launder BE, Sandsam ND (eds) Closure strategies for turbulent and transitional flows. Cambridge University Press, Cambridge, pp 464–492

Seyfert C, Krumbein A (2012) Correlation-based transition transport modeling for three-dimensional aerodynamic configuration. AIAA paper 2012-0448. Paper presented at 50th AIAA Aerospace Sciences Meeting. Nashville, Tennessee, 9–12 Jan 2012

Sheng C (2014) Predictions of JVX rotor performance in hover and airplane mode using high-fidelity unstructured grid CFD solver. In: Proceedings of the AHS 70th annual forum, Montreal, Canada, 20–22 May 2014

Sheng C, Wang J, Zhao Q (2016) Improved rotor hover predictions using advanced turbulence modeling. J Aircraft. doi:10.2514/1.C033512

Smagorinsky J (1963) General circulation experiments with the primitive equations. I: the basic experiment. Monthly Weather Rev 91(3):99–164

Smith AMO, Gamberoni N (1956) Transition, pressure gradient and stability theory. Douglas Aircraft Co. Rept. ES26388, El Segundo, California

Spalart PR, Allmaras SR (1994) A one-equation turbulence model for aerodynamic flows. La Recherche Aerospatiale 1994(1):5–21

Steeland J, Dick E (2001) Modeling of laminar—turbulent transition for high freestream turbulence. J Fluid Mech 123(1):22–30

Stock HW, Haase W (2000) Navier-Stokes airfoil computations with e^N transition prediction including transitional flow regions. AIAA J 38(11):2059–2066

Suzen YB, Huang PG (2000) Modeling of flow transition using an intermittency transport equation. ASME J Fluids Eng 122(2):273–284

Suzen YB, Xiong G, Huang PG (2002) Predictions of transitional flows in low-pressure turbines using intermittency transport equation. AIAA J 40(2):254–266

Sveningsson A (2006) Transition modelling—a review. Technical report, Chalmers University of Technology, Gothenburg, Sweden

van Ingen JL (1956) A suggested semi-empirical method for the calculation of the boundary layer transition region. University of Technology, Department of Aerospace Engineering, Report UTH-74, Delft

Volino RJ (1998) A new model for free-stream turbulence effects on boundary layers. ASME J Turbomach 120:613–620

Walters DK, Leylek JH (2002) A new model for boundary-layer transition using a single-point RANS approach. In: ASME IMECE'02, IMECE2002-HT-32740

Walters DK, Leylek JH (2004) A new model for boundary-layer transition using a single-point RANS approach. J Turbomach 126(1):193–202. doi:10.1115/1.1622709

Wang J, Sheng C (2014) Validations of a local correlation-based transition model using an unstructured grid CFD solver. AIAA-2014-2211. Paper presented at 7th AIAA Theoretical Fluid Mechanics Conference, Atlanta, Georgia, 16–20 June 2014

Wang J, Sheng C (2015) A comparison of a local correlation-based transition model coupled with SA and SST turbulence models. AIAA-2015-0587. Paper presented at 53rd AIAA Aerospace Sciences Meeting, Kissimmee, Florida, 5–9 Jan 2015

Warren EW, Hassan HA (1997) Alternative to the e^N method for determining onset of transition. AIAA J 36(1):111–113

Warren ES, Harris JE, Hassan HA (1995) Transition model for high-speed flow. AIAA J 33 (8):1391–1397

Wilcox D (1992) Dilatation-dissipation corrections for advanced turbulence models. AIAA J 30 (11):2639–2646

Wu X, Durbin PA (2001) Evidence of longitudinal vortices evolved from distorted wakes in a turbine passage. J Fluid Mech 446:199–228

Wu X, Jacobs RG, Hunt JCR et al (1999) Simulation of boundary layer transition induced by periodically passing wakes. J. Fluid Mech 398:109–153

Chapter 3
Transition Model

Abstract A Local Correlation-based Transition Model is described in this chapter. This single-point transition model is integrated into both the Spalart-Allmaras turbulence model and Menter's Shear Stress Transport turbulence model. A method to specify the local free-stream turbulence intensity is presented for use in the Spalart-Allmaras transition model. A new separation correction, called stall delay method, is introduced in this chapter in order to correct the model's behavior near separation or reattachment points. Numerical procedures are described in detail based on an implicit Newton's method for solving the transport equations.

3.1 The Langtry-Menter Transition Model

Menter et al. (2002, 2004, 2006a, b), as well as Langtry and Menter (2006, 2009), proposed a single-point Local Correlation-Based Transition Model called $\gamma - \widetilde{Re}_{\theta t}$. The central idea behind this method is to use a concept of vorticity Reynolds number (Re_v) proposed by Driest and Blumer (1963), which links the local transition onset Reynolds number ($Re_{\theta t}$) and the boundary layer velocity profile. Because the vorticity Reynolds number is a local variable, integration of the boundary layer profile for $Re_{\theta t}$ is thus avoided. Combined with empirical correlations for the transition onset Reynolds number, this method can be easily incorporated into general structured and unstructured grid CFD codes.

The vorticity Reynolds number (or alternatively the strain rate Reynolds number) is defined as:

$$Re_v = \frac{\rho y^2}{\mu} \left| \frac{\partial u}{\partial y} \right| = \frac{\rho y^2}{\mu} S \tag{3.1}$$

where ρ is density, μ is dynamic viscosity, u is the local velocity, y is the distance to the nearest wall and S is the shear strain rate. Examining the Blasius boundary layer velocity profile, Driest and Blumer (1963) concluded that there is a limiting value of the vorticity Reynolds number in an undisturbed Blasius boundary layer, and the

© The Author(s) 2017
C. Sheng, *Advances in Transitional Flow Modeling*, SpringerBriefs
in Applied Sciences and Technology, DOI 10.1007/978-3-319-32576-7_3

initial laminar breakdown should occur at the location that coincides with the maximum value of the vorticity Reynolds number. Langtry and Sjolander (2002) further found that the ratio of the maximum vorticity Reynolds number to the momentum thickness Reynolds number (Re_θ) holds a constant value of 2.193, at least for bypass transitions:

$$\frac{\max(Re_v)}{Re_\theta} = 2.193 \tag{3.2}$$

According to Driest and Blumer (1963), the maximum value of the vorticity Reynolds number occurs at $y/\delta = 0.57$ (or $\eta = 2.015$) in the Blasius boundary layer (White 2006). The distribution of the ratio between the vorticity Reynolds number and the momentum thickness Reynolds number in the boundary layer can be drawn as Fig. 3.1.

It should be noted that the maximum value for the Re_v/Re_θ in Fig. 3.1 may be considered as a criterion of transition in the Blasius boundary layer with a zero-pressure gradient. However, pressure gradients can change the boundary layer velocity profile and the shape factor (H), and then affect the ratio between the maximum vorticity Reynolds number and the momentum thickness Reynolds number. For moderate pressure gradients (2.3 < H < 2.9) where the majority of experimental data are obtained from, the difference between the maximum vorticity Reynolds number and the momentum thickness Reynolds number is within 10 % (Menter et al. 2002). However, for strong adverse pressure gradients, the difference between these two Reynolds numbers becomes significant, especially near the separation point (H = 3.5), which means that the criterion of transition has to be modified.

Two transport equations were proposed by Menter et al. (2002, 2004, 2006a, b): one for intermittency (γ) and another for the transition onset momentum thickness Reynolds number ($Re_{\theta t}$). The equation of intermittency is solved to distinguish the

Fig. 3.1 Ratio of vorticity and momentum thickness Reynolds numbers versus non-dimensional distance in a Blasius boundary layer

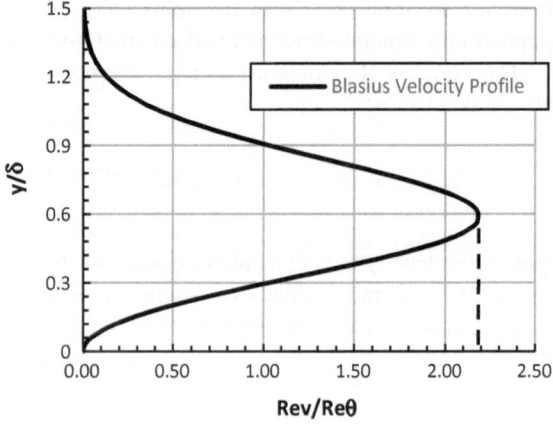

laminar and turbulent regions during the transition process, which has a value of zero for laminar flow and unity for fully turbulent flow. The equation for the transition onset momentum thickness Reynolds number, formulated as a scalar quantity, $\widetilde{Re}_{\theta t}$, is solved in order to capture non-local influences of the turbulence intensity and the change of the free-stream velocity outside the boundary layer. The second transport equation is essential to this transition model, as it builds the link between the empirical correlation to the onset criteria in the intermittency equation. The advantage of using combined transport equations and an empirical correlation is the separation of viscous sublayer damping and transition prediction, which allows calibration of the models without affecting the physics of turbulence (Menter et al. 2002, 2004, 2006a, b).

In the following sections, the $\gamma - \widetilde{Re}_{\theta t}$ transition model proposed by Langtry and Menter (2009) is described. Modifications to the original model are introduced in order to integrate it into the Spalart-Allmaras (S-A) turbulence model and to correct the model's behavior near separation points. Integration of the $\gamma - \widetilde{Re}_{\theta t}$ model with the S-A turbulence model yields a three-equation SA-based transition model (SA-TM), and integration with the SST model yields a four-equation SST-based transition model (SST-TM). General procedures are provided for coupling the transition model with both turbulence models, including the boundary conditions recommended by Spalart and Rumsey (2007) for external aerodynamic calculations. Numerical procedures are detailed for solving the SA-TM and SST-TM transition models using a modern implicit up-winding scheme.

3.1.1 Transport Equation of Intermittency

The first transport equation in the $\gamma - \widetilde{Re}_{\theta t}$ model is for intermittency (γ), which is expressed by Menter et al. (2002, 2004, 2006a, b) as:

$$\frac{\partial(\rho\gamma)}{\partial t} + u_j \frac{\partial(\rho\gamma)}{\partial x_j} = P_\gamma - E_\gamma + \frac{\partial}{\partial x_j}\left[\left(\mu + \frac{\mu_t}{\sigma_\gamma}\right)\frac{\partial\gamma}{\partial x_j}\right] \tag{3.3}$$

where ρ is density, μ is the molecular viscosity, μ_t is the eddy viscosity and σ_r is a constant for the intermittency equation. P_γ and E_γ are the sources of the production and destruction terms, respectively. For the intermittency equation, P_γ and E_γ are expressed as:

$$P_\gamma = F_{length}c_{a1}\rho S[\gamma F_{onset}]^{0.5}(1 - c_{e1}\gamma) \tag{3.4}$$

$$E_\gamma = c_{a2}\rho\Omega\gamma F_{turb}(c_{e2}\gamma - 1) \tag{3.5}$$

In the above expressions, S is the magnitude of the strain rate, and Ω is the magnitude of the vorticity. F_{length} and F_{onset} are the two important parameters that control the production of intermittency. F_{length} is used to control the length of the transition region and is based on an empirical correlation provided by Langtry and Menter (2009). There is a strong relationship between F_{length} and the transition onset momentum thickness Reynolds number, which will be described in Sect. 3.1.3. The second important parameter is F_{onset}, which is used to trigger the production of intermittency when the local vorticity Reynolds number (Re_v) exceeds the local transition onset criteria. The function for F_{onset} is defined as:

$$F_{onset1} = \frac{Re_v}{2.193 \cdot Re_{\theta c}} \tag{3.6a}$$

$$F_{onset2} = min\left(max\left(F_{onset1}, F_{onset1}{}^4\right), 2.0\right) \tag{3.6b}$$

$$F_{onset3} = max\left(1 - \left(\frac{R_T}{2.5}\right)^3, 0\right) \tag{3.6c}$$

$$F_{onset} = max(F_{onset2} - F_{onset3}, 0) \tag{3.6d}$$

$$F_{onset} = max(F_{onset2} - F_{onset3}, 0) \tag{3.6e}$$

where Re_v is the vorticity Reynolds number defined in Eq. (3.1). $Re_{\theta c}$ is the critical momentum thickness Reynolds number where turbulence or intermittency starts to grow in the boundary layer. The value of $Re_{\theta c}$ can be obtained by an empirical correlation, expressed as a low order polynomial of the local transition onset momentum thickness Reynolds number, which is obtained from the second transport equation. The empirical correlation for $Re_{\theta c}$ will be given in Sect. 3.1.3. R_T in Eq. (3.6c) is the ratio of the eddy viscosity and the molecular viscosity, which is expressed in the following two forms. For the Spalart-Allmaras (S-A) turbulence model,

$$R_T = \frac{\mu_t}{\mu} \tag{3.7a}$$

and for the Menter Shear Stress Transport (SST) turbulence model,

$$R_T = \frac{\rho k}{\mu \omega} \tag{3.7b}$$

The destruction source term, E_γ, serves to ensure that the intermittency remains close to zero in the laminar boundary layer. E_γ is controlled by a parameter F_{turb}, which is defined by:

$$F_{turb} = e^{-\left(\frac{R_T}{4.0}\right)^4} \tag{3.8}$$

The constant c_{a2} in Eq. (3.5) is used to control the magnitude of the destruction term, which ensures that the production term (P_γ) is larger than the destruction term (E_γ). All constants used in the above expressions are given as:

$$\sigma_\gamma = 1.0; \quad c_{a1} = 2.0; \quad c_{a2} = 0.06; \quad c_{e1} = 1.0; \quad c_{e2} = 50.0 \qquad (3.9)$$

The boundary condition for the intermittency equation is a zero normal flux at the wall and a value of unity at the free-stream boundary. The boundary condition for R_T is a zero value at the wall and the ratio of the ambient eddy viscosity and the molecular viscosity at the free-stream boundary.

3.1.2 Transport Equation of Transition Onset Reynolds Number

In order to calculate the critical Reynolds number $(Re_{\theta c})$ in Eq. (3.6a) and to include the non-local influence of the turbulence intensity, a second transport equation is required to solve the local transition onset momentum thickness Reynolds number $(Re_{\theta t})$, which is formulated in terms of the scalar quantity, $\widetilde{Re}_{\theta t}$. The transport equation for $\widetilde{Re}_{\theta t}$ is written as:

$$\frac{\partial(\rho \widetilde{Re}_{\theta t})}{\partial t} + u_j \frac{\partial(\rho \widetilde{Re}_{\theta t})}{\partial x_j} = P_{\theta t} + \frac{\partial}{\partial x_j}\left[\sigma_{\theta t}(\mu + \mu_t)\frac{\partial}{\partial x_j}\widetilde{Re}_{\theta t}\right] \qquad (3.10)$$

According to Langtry and Menter (2006) and Langtry et al. (2006), the production term $(P_{\theta t})$ forces the transported scalar $\widetilde{Re}_{\theta t}$ to match the local value of $Re_{\theta t}$ calculated from the empirical correlation outside of the boundary layer. $P_{\theta t}$ is turned off in the boundary layer to allow the transported scalar $\widetilde{Re}_{\theta t}$ to diffuse from the free stream. The source of the production term $(P_{\theta t})$ can be written as:

$$P_{\theta t} = c_{\theta t}\frac{\rho}{t}\left(Re_{\theta t} - \widetilde{Re}_{\theta t}\right)(1 - F_{\theta t}) \qquad (3.11)$$

where t is a time scale that is constructed for dimensional reasons. The value of t is defined as:

$$t = \frac{500\mu}{\rho U^2} \qquad (3.12)$$

$F_{\theta t}$ is a blending function to turn off the source term outside the boundary layer, which is zero in the free stream and one inside the boundary layer. $F_{\theta t}$ can be expressed as:

$$F_{\theta t} = min\left(max\left(F_{wake}e^{-\left(\frac{y}{\delta}\right)^4}, 1.0 - \left(\frac{\gamma - 1/c_{e2}}{1.0 - 1/c_{e2}}\right)^2\right), 1.0\right) \qquad (3.13a)$$

$$F_{wake} = e^{-\left(\frac{Re_\omega}{10^5}\right)^2} \qquad (3.13b)$$

A few of variables in Eqs. (3.13a) and (3.13b) are defined as:

$$Re_\omega = \frac{\rho y^2}{\mu}\omega \qquad (3.14a)$$

$$\delta = \frac{50\Omega y}{U}\theta_{BL}, \quad \delta_{BL} = \frac{15}{2}\theta_{BL}, \quad \theta_{BL} = \frac{Re_{\theta t}\mu}{\rho U} \qquad (3.14b, c, d)$$

Two constants used in the above expressions are:

$$\sigma_{\theta t} = 2.0; \; c_{\theta t} = 0.03 \qquad (3.15)$$

The boundary condition for $\widetilde{Re}_{\theta t}$ is a zero normal flux at the wall, and is calculated based on the empirical correlation of the turbulence intensity at the free-stream boundary.

3.1.3 Correlation Formula

The transport equations described for γ and $\widetilde{Re}_{\theta t}$ in the preceding sections cannot be solved without providing the closure equations for F_{onset}, F_{length}, and $Re_{\theta t}$. To control the onset and the length of transition in the $\gamma - \widetilde{Re}_{\theta t}$ model, F_{onset} and F_{length} are used. F_{onset} is determined by a critical Reynolds number ($Re_{\theta c}$), which is the value where the intermittency first starts to increase in the boundary layer. From the flow physics standpoint, it is reasonable to assume that $Re_{\theta c}$ occurs upstream of the transition onset location. In other words, $Re_{\theta c}$ must not exceed $Re_{\theta t}$.

Before the disclosure of the Langtry and Menter's empirical correlations, Suluksna et al. (2009) proposed a linear relationship to relate $Re_{\theta c}$ with $\widetilde{Re}_{\theta t}$, where $Re_{\theta c} = \alpha_{\theta t}\widetilde{Re}_{\theta t}$ and $\alpha_{\theta t} = 0.8$. This linear relationship was later adopted by Medida and Baeder (2011) to couple the $\gamma - \widetilde{Re}_{\theta t}$ model into the S-A turbulence model. The impact of the constant value of $\alpha_{\theta t}$ on the transition onset location is illustrated in Fig. 3.2 for a flat plate with a zero-pressure gradient, where the $\gamma - \widetilde{Re}_{\theta t}$ transition model was solved with the S-A turbulence model (Wang and Sheng 2014). For fixed values of F_{length} and the free-stream turbulence intensity (Tu), an increase in $Re_{\theta c}$ decreases the F_{onset} parameter, which causes a delay of the transition onset.

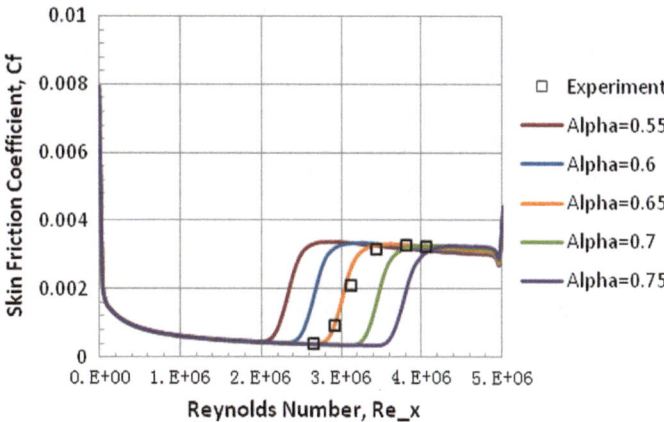

Fig. 3.2 Comparison of skin friction profiles for α on flat plate

Likewise, a decrease in $Re_{\theta c}$ increases the F_{onset} parameter, which results in an early transition.

The value of F_{length} also has a large impact on the transition process, *i.e.*, the length of the transition. The explicit form of the function F_{length} is complicated, which varies with the velocity profile shape. Suluksna et al. (2009) suggested a value of 100 as a starting point for F_{length}. The influence of F_{length} on the transition length is illustrated in Fig. 3.3 for the same flat plate as above, where the values of F_{length} are normalized by the reference Reynolds number. At the same transition onset location, an increased value of F_{length} will increase the production of inter- mittency and turbulence, which causes the rapid growth rate of the transition

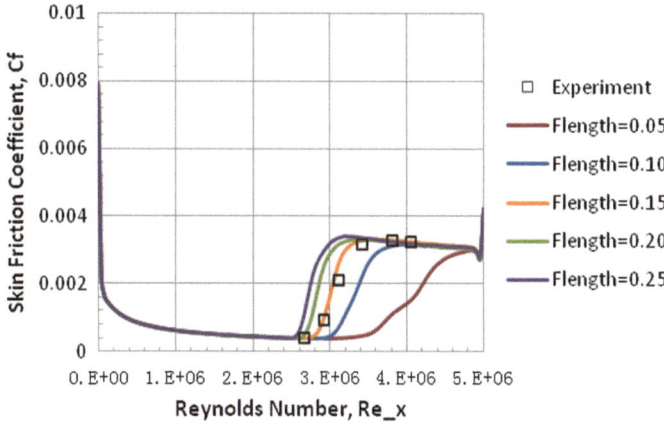

Fig. 3.3 Comparison of skin friction profiles for F_{length} on flat plate

leading to a shorter transition length. On the other hand, a decrease in F_{length} will reduce the turbulence production, and prolongs the process of the transition.

Langtry and Menter (2009) later published their empirical correlations to calculate $Re_{\theta c}$ and F_{length}, which were calibrated based on a wide range of cases from simple two-dimensional airfoils to complex engineering applications. The empirical correlations for $Re_{\theta c}$ and F_{length} are defined as polynomial functions of $\widetilde{Re}_{\theta t}$, which are given as:

$$
Re_{\theta c}
$$

$$
= \begin{cases}
[\widetilde{Re}_{\theta t} - (396.035 \times 10^{-2} - 120.656 \times 10^{-4} \cdot \widetilde{Re}_{\theta t} \\
\quad -868.230 \times 10^{-6} \cdot \widetilde{Re}_{\theta t}^2 + 696.506 \times 10^{-9} \cdot \widetilde{Re}_{\theta t}^3 + \quad \widetilde{Re}_{\theta t} \leq 1870 \\
\quad\quad\quad 174.105 \times 10^{-12} \cdot \widetilde{Re}_{\theta t}^4)] \\
[\widetilde{Re}_{\theta t} - (593.11 + (\widetilde{Re}_{\theta t} - 1870.0) \cdot 0.48)] \quad \widetilde{Re}_{\theta t} \geq 1870
\end{cases}
$$

$$(3.16)$$

$$
F_{length}
$$

$$
= \begin{cases}
[398.189 \times 10^{-1} - 119.270 \times 10^{-4} \cdot \widetilde{Re}_{\theta t} \\
\quad -868.230 \times 10^{-6} \cdot \widetilde{Re}_{\theta t}^2] & \widetilde{Re}_{\theta t} < 400 \\
[263.404 - 123.939 \times 10^{-2} \cdot \widetilde{Re}_{\theta t} + \\
\quad 194.548 \times 10^{-5} \cdot \widetilde{Re}_{\theta t}^2 - & 400 \leq \widetilde{Re}_{\theta t} < 596 \\
\quad 101.695 \times 10^{-8} \cdot \widetilde{Re}_{\theta t}^3] \\
[0.5 - (\widetilde{Re}_{\theta t} - 596.0) \cdot 3.0 \times 10^{-4}] & 596 \leq \widetilde{Re}_{\theta t} < 1200 \\
[0.3188] & 1200 \leq \widetilde{Re}_{\theta t}
\end{cases}
$$

$$(3.17)$$

For the transport equation of the momentum thickness Reynolds number ($Re_{\theta t}$), an empirical correlation was also developed for both zero and non-zero pressure gradient flows. The empirical correlation developed by Langtry and Menter (2009) is similar to that of Abu-Ghannam and Shaw (1980), which is a function of both the pressure gradient parameter (λ_θ) and the free-stream turbulence intensity (Tu):

$$
Re_{\theta t} = f(\lambda_\theta, Tu)
$$

$$
= \begin{cases}
\left(1173.51 - 589.428 \cdot Tu + \frac{0.2196}{Tu^2}\right) F(\lambda_\theta), & Tu \leq 1.3 \\
331.50(Tu - 0.5658)^{-0.671} F(\lambda_\theta), & Tu > 1.3
\end{cases}
$$

$$(3.18a)$$

$$
F(\lambda_\theta)
$$

$$
= \begin{cases}
1 - \left(-12.986\lambda_\theta - 123.66\lambda_\theta^2 - 405.689\lambda_\theta^3\right) e^{-\left[\frac{Tu}{1.5}\right]^{1.5}}, & \lambda_\theta \leq 0 \\
1 + 0.275\left(1 - e^{-35.0\lambda_\theta}\right) e^{-\frac{Tu}{0.5}}, & \lambda_\theta > 0
\end{cases}
$$

$$(3.18b)$$

where the momentum thickness Reynolds number, the pressure gradient parameter and the turbulence intensity are defined as:

$$Re_{\theta t} = \frac{\rho U \theta}{\mu} = f(\lambda_\theta, Tu) \tag{3.19}$$

$$\lambda_\theta = \frac{\rho \theta^2}{\mu} \frac{dU}{ds} \tag{3.20}$$

$$Tu = 100 \frac{\sqrt{2k/3}}{U} \tag{3.21}$$

where dU/ds is the acceleration along the stream-wise direction, which can be computed by taking the derivative of the total velocity (U) in the x, y and z directions and then summing the contribution along the stream-wise flow direction:

$$U = \left(u^2 + v^2 + w^2\right)^{\frac{1}{2}} \tag{3.22a}$$

$$\frac{dU}{dx} = \frac{1}{2}\left(u^2 + v^2 + w^2\right)^{-\frac{1}{2}} \cdot \left[2u\frac{du}{dx} + 2v\frac{dv}{dx} + 2w\frac{dw}{dx}\right] \tag{3.22b}$$

$$\frac{dU}{dy} = \frac{1}{2}\left(u^2 + v^2 + w^2\right)^{-\frac{1}{2}} \cdot \left[2u\frac{du}{dy} + 2v\frac{dv}{dy} + 2w\frac{dw}{dy}\right] \tag{3.22c}$$

$$\frac{dU}{dz} = \frac{1}{2}\left(u^2 + v^2 + w^2\right)^{-\frac{1}{2}} \cdot \left[2u\frac{du}{dz} + 2v\frac{dv}{dz} + 2w\frac{dw}{dz}\right] \tag{3.22d}$$

$$\frac{dU}{ds} = \left[\frac{u}{U}\frac{dU}{dx} + \frac{v}{U}\frac{dU}{dy} + \frac{w}{U}\frac{dU}{dz}\right] \tag{3.22e}$$

Recall that both the pressure gradient parameter in Eq. (3.20) and the turbulence intensity in Eq. (3.21) require the total fluid velocity and its gradients being evaluated with respect to stationary walls in the coordinate system, which are not Galilean invariants (Menter et al. 2015). For problems involving moving walls such as turbomachinery blades and helicopter rotors, a relative velocity and its gradients with respect to the moving surfaces should be evaluated in the above expressions. The above empirical correlations are used in the source term of the transport equation for the transition onset Reynolds number, which has to be solved iteratively because the momentum thickness (θ) is an unknown in the equation. Newton's method can be used to solve the equation, with an initial guess of the pressure gradient parameter λ_θ to be zero as follows:

$$F(\theta) = \frac{\rho U \theta}{\mu} - Re_{\theta t}(\lambda_\theta, Tu) \tag{3.23a}$$

$$F'\left(\theta^{n+1}\right)\left(\theta^{n+1} - \theta^n\right) = -F(\theta^n) \tag{3.23b}$$

$$\theta^{n+1} = \theta^n - \frac{F(\theta^n)}{F'\left(\theta^{n+1}\right)} \tag{3.23c}$$

where n is Newton's iteration step. In general, eight to ten Newton's iterations are sufficient to converge the solution of the momentum thickness (θ) to a numerically stable state. The value of variables in the empirical correlation should be limited to avoid an unstable solution during the iteration. The suggested ranges are as follows:

$$-0.1 \leq \lambda_\theta \leq 0.1 \tag{3.24a}$$

$$Tu \geq 0.027\,\% \tag{3.24b}$$

$$Re_{\theta t} \geq 20 \tag{3.24c}$$

3.2 Modifications of Transition Model

The transition model proposed by Menter et al. (2002, 2004, 2006a, b) was originally developed under the framework of the SST k-ω two-equation turbulence model (Menter 1994). Later, Medida and Baeder (2011, 2013) extended it into the widely used Spalart-Allmaras (S-A) one-equation turbulence model (Spalart and Allmaras 1994). While the majority of the formulations in the transition model remain the same, a major difference exists in the way to determine the local turbulence intensity (Tu). In the original SST implementation, Tu is calculated from the local kinetic energy term. However, in the S-A model, this information is not available since the model solves for the eddy viscosity directly. Medida and Baeder (2011) proposed a constant local turbulence intensity based on the value at the inlet or free stream. Wang and Sheng (2014) proposed an improved way to determine the local turbulence intensity by considering the decay of the turbulence from the free-stream value. Effects of different treatments of the local turbulence intensity on the transition model will be discussed in this section. In addition, a new method to correct the model's behavior close to separation points, called the Stall Delay Method (SDM), is introduced here in order to prevent numerically induced flow separations in the rotor performance predictions. This modification is essential for capturing the flow physics in rotatory wings and obtaining accurate predictions of aerodynamic performance for the entire operating envelope (Sheng et al. 2016).

3.2.1 *Local Free-Stream Turbulence Intensity*

From Eq. (3.16), the transition onset Reynolds number is determined based on both turbulence intensity and pressure gradient. A decrease in turbulence intensity would increase the transition onset Reynolds number, and thus move the transition onset location downstream. On the contrary, an increase in turbulence intensity would decrease the transition onset Reynolds number and move the transition onset location upstream. The comparison of the skin friction profiles obtained on a flat plate at a zero-pressure gradient, as shown in Fig. 3.4, clearly shows the impact of the free-stream turbulence intensity on the onset of transition. When the level of free-stream turbulence intensity (Tu_∞) is very low, such as less than 0.01 %, the boundary layer flow is laminar on the entire flat plate. With an increased value of Tu_∞, a natural transition onset occurs at different locations on the flat plate.

In the original $\gamma - \widetilde{Re}_{\theta t}$ model, the turbulence intensity is a local variable and is calculated directly from the turbulent kinetic energy term in the SST κ-ω turbulence model (Menter 1994). However, since the S-A turbulence model solves the transport equation for the eddy viscosity directly, it cannot provide the information about local turbulence intensity. Medida and Baeder (2011) proposed to use a far field constant value as the local turbulence intensity in the coupling of the $\gamma - \widetilde{Re}_{\theta t}$ model and the S-A turbulence model. This method, however, may cause incorrect local turbulence level since the turbulence free decay rate varies widely depending on the inlet or far field conditions. Spalart and Rumsey (2007) performed a numerical study about the inflow conditions for the S-A and SST turbulence models in aerodynamic calculations. They found that either a higher level of free-stream turbulence intensity or a lower level of turbulent eddy viscosity would cause a rapid decay of turbulence in the field. In particular, their numerical results showed a very

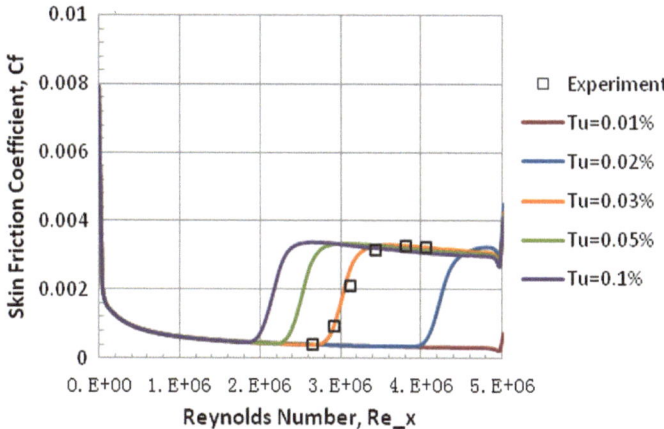

Fig. 3.4 Comparison of skin friction profiles using different free stream turbulence intensity on a flat plate

noticeable grid dependence for the higher free-stream decay condition (Spalart and Rumsey 2007). This indicates that a correct inflow condition is required in order to achieve a physically meaningful solution in aerodynamic calculations. In order to provide a more realistic estimate of the local turbulence level, Wang and Sheng (2015) proposed a new approach to estimate the local free-stream turbulence intensity by considering the decay of the free-stream turbulence from the inlet or far field boundary.

Recall that in the SST turbulence model, the dissipation of the turbulent kinetic energy is realized through the destruction term. One may consider the transport equations of kinetic energy (k) and turbulence frequency (ω) without the production and cross-diffusion terms as:

$$\frac{Dk}{Dt} = -\beta^* \rho k \omega \tag{3.25}$$

$$\frac{D\omega}{Dt} = -\beta \rho \omega^2 \tag{3.26}$$

where β^* and β are two constants in the SST turbulence model, and ρ is density. The above two equations are solved with the following initial conditions:

$$k = k_\infty; \quad \omega = \omega_\infty \text{ at } t = 0, \tag{3.27}$$

The derived solutions can then be written as:

$$k = k_\infty (1 + \omega_\infty \beta t)^{-\frac{\beta^*}{\beta}} \tag{3.28}$$

$$\omega = \omega_\infty / (1 + \omega_\infty \beta t) \tag{3.29}$$

Assuming that the free stream has a velocity of U_∞, and x is the distance from the inlet or far-field boundary, the time term t can then be written as $t = x/U_\infty$. With additional definitions of the viscosity ratio (R_T) and the turbulence intensity (Tu) defined in Eqs. (3.7b) and (3.21), the local free-stream turbulence intensity at the transition onset location will be decayed from the value at the inlet or far-field boundary based on the following expression:

$$Tu = Tu_\infty (1 + \frac{3}{2} \beta Re_x \frac{Tu_\infty^2}{R_{T\infty}})^{-\frac{\beta^*}{2\beta}} \tag{3.30}$$

where Re_x is the local Reynolds number, and x is the distance from the transition onset location to the inlet or far-field boundary. β and β^* are constants used in the SST k-ω model.

The expression (3.30) for the local free-stream turbulence intensity is evaluated here based on the calculations of the European Research Community on Flow, Turbulence and Combustion (ERCOFTAC) T3 series of experimental tests on flat

Table 3.1 Turbulence intensities on a flat plate

Test case	T3A	T3B	T3AM	S&K	T3C2	T3C3	T3C4	T3C5
Inlet free-stream Tu (%)	3.3	6.5	0.874	0.3	3.0	3.0	3.0	4.0
Local free-stream Tu (%)	1.89	5.72	0.545	0.154	1.56	1.39	1.0	2.24

plates. Table 3.1 shows the free-stream Tu values at both inlet and the test section for various test cases. The decayed values at the flat plate location are different from that specified at the inlet. However, evaluating the local free-stream turbulence intensity using Eq. (3.30) needs the distance from the inlet to the flat plate, which is to ensure that the local free-stream turbulence intensity in the field matches the measurement in the wind tunnel test. Without such information, a constant ambient value recommended by Spalart and Rumsey (2007) is used in the S-A transition model, which will be further discussed in Sect. 3.3.

Figures 3.5, 3.6, 3.7 and 3.8 show the effects of using the fixed inlet free-stream turbulence intensity and the local value evaluated using Eq. (3.30). The skin friction coefficients were calculated over the flat plat at a zero-pressure gradient. Using the constant free-stream turbulence intensity method, the transition onset location was prematurely predicted when compared to the experiment. This is because that un-decayed free-stream turbulence value results in a lower critical Reynolds number for the transition onset criterion as described in Eq. (3.16). The local free-turbulence intensity based on Eq. (3.30) provided a more realistic estimate of the free-stream turbulence intensity at the flat plate location, and thus a more accurate transition onset criterion as shown in these figures.

Similar effects are also demonstrated for the flat plate under the influence of pressure gradients in Figs. 3.9, 3.10, 3.11 and 3.12. Like the results obtained at the

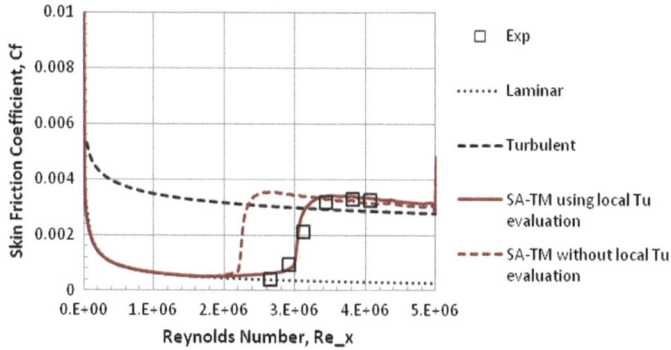

Fig. 3.5 Comparison of the skin friction profiles using different turbulence intensities for the S&K flat plate

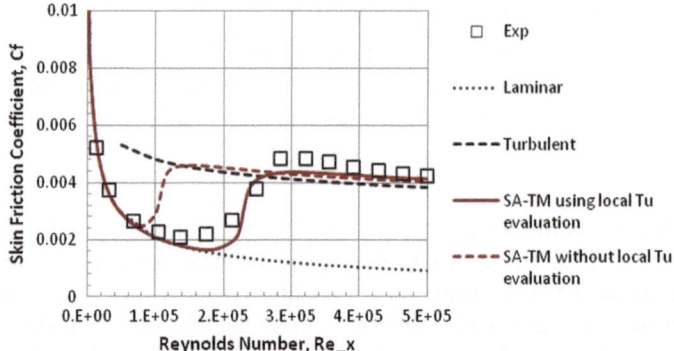

Fig. 3.6 Comparison of the skin friction profiles using different turbulence intensities for the T3A flat plate

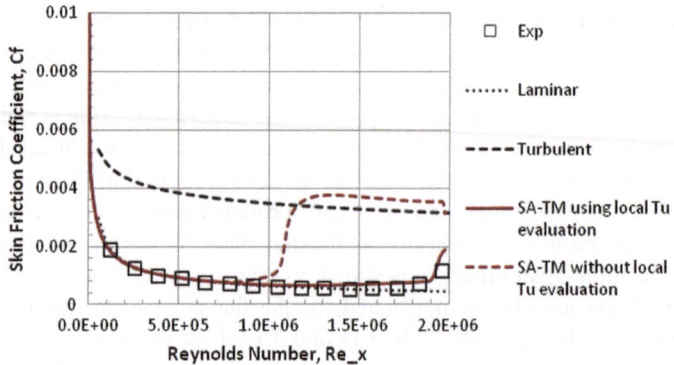

Fig. 3.7 Comparison of the skin friction profiles using different turbulence intensities for the T3AM flat plate

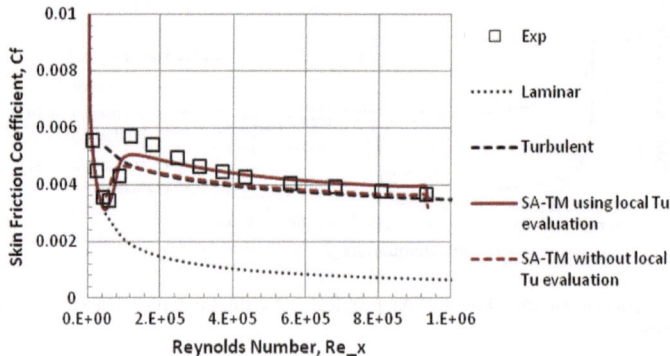

Fig. 3.8 Comparison of the skin friction profiles using different turbulence intensities for the T3B flat plate

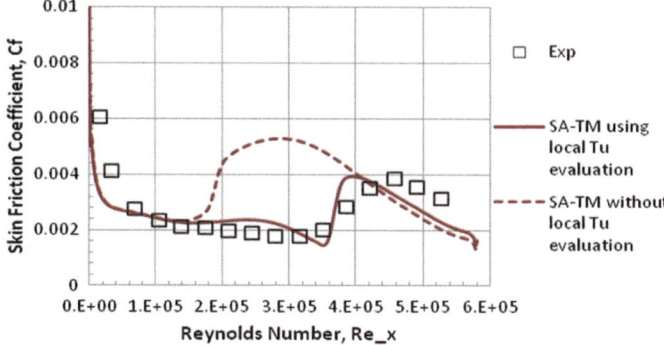

Fig. 3.9 Comparison of the skin friction profiles using different turbulence intensities for the T3C2 flat plate

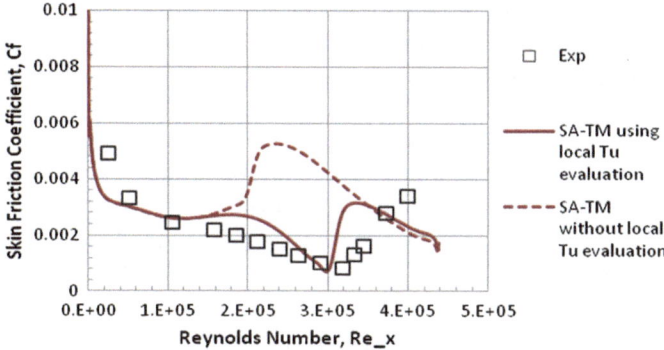

Fig. 3.10 Comparison of the skin friction profiles using different turbulence intensities for the T3C3 flat plate

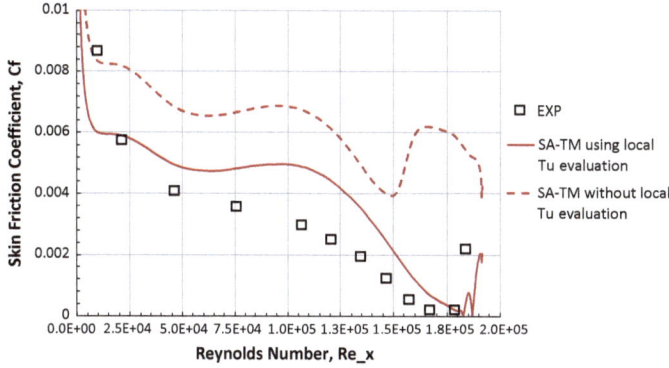

Fig. 3.11 Comparison of the skin friction profiles using different turbulence intensities for the T3C4 flat plate

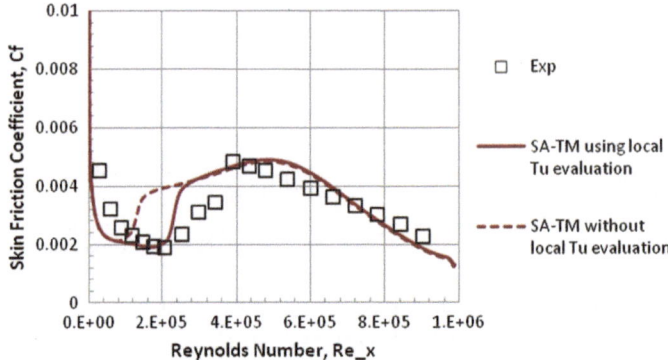

Fig. 3.12 Comparison of the skin friction profiles using different turbulence intensities for the T3C5 flat plate

zero-pressure gradient, the constant free-stream turbulence intensity method was unable to predict the correct transition onset location, especially for the T3C4 case where no separation bubble was even captured. The local free-stream turbulence intensity method significantly improved the numerical accuracy in predicting the transition onset location on flat plates with the pressure gradients. A slightly higher C_f predicted in CFD indicated a mismatched local free-stream velocity or pressure gradient distribution compared with the experiment, which will be further discussed in Chap. 4.

3.2.2   Separation Correction Method

In the development of the Langtry-Menter transition model, Langtry (2006) discovered that the model has consistently predicted the turbulent reattachment location to be too far downstream whenever a laminar boundary layer separation occurred. He presumed that reduced turbulent kinetic energy (k) predicted in the separating shear layer caused a lower level of free-stream turbulence intensity, resulting in over-prediction of the laminar separation bubble. To correct this deficiency, he proposed a modification to the transition model by allowing the kinetic energy to grow rapidly once the laminar boundary layer separates. Since the vorticity Reynolds number (Re_v) significantly exceeds the critical momentum thickness Reynolds number ($Re_{\theta c}$) within the separated flows, the ratio between the two Reynolds numbers can be viewed as a measure of the size of the laminar separation. The following modified intermittency, proposed by Langtry (2006), served to increase the turbulent kinetic energy in order to predict separation-induced transition:

$$\gamma_{sep} = \min\left(s_1 \max\left[0, \frac{Re_v}{3.235 \cdot Re_{\theta c}} - 1 \right] F_{reattach}, 2.0 \right) \cdot F_{\theta t} \qquad (3.31a)$$

$$F_{reattach} = e^{-\left(\frac{R_t}{20}\right)^4} \qquad (3.31b)$$

$$\gamma_{eff} = \max\left(\gamma, \gamma_{sep}\right) \qquad (3.31c)$$

$$s_1 = 2.0 \qquad (3.31d)$$

In the above expressions, $F_{reattach}$ is a function to disable the modification once the viscosity ratio is large enough to ensure reattachment. $F_{\theta t}$ is the blending function from the transport equation (3.12) to confine the modification to the boundary layer flow. The ratio of 3.235 between Re_v and $Re_{\theta c}$ corresponds to a shape factor (H) of 3.5 for a separated velocity profile.

It should be noted that the above modification to the turbulent kinetic energy is only made within the transition flow region, which does not affect the physics in fully turbulent flows (Langtry 2006). However, there is a well-known issue faced by the RANS modelling approaches, which is that they generally underestimate the turbulence when flows are close to the separation or reattachment points (Aupoix et al. 2011). The underestimate of turbulence (kinetic energy or eddy viscosity) in turbulent flows may cause a larger separation zone than in reality, which was the issue encountered by the author in predicting the aerodynamic performance for helicopter and tilt rotors operating at high thrusts (Sheng 2014; Sheng et al. 2016). In fact, this problem was also reported by several other researchers (Jung et al. 2014; Gardarein and Le Pape 2016; Min and Wake 2016) in predicting a conventional helicopter rotor performance at high thrusts. An enlarged flow separation in the blade tip region causes deviation of predicted rotor performance from the experimental value. Numerical experiments using the Langtry's separation correction (3.31) appeared to not be sufficient to correct this deficiency, which indicates that it may be necessary to modify the behavior of turbulence models near separations.

In order to address the deficiency of the RANS modeling and improve the aerodynamic prediction especially for rotors near separations, a new separation correction method was introduced (Sheng 2014; Sheng et al. 2016), which removes the blending function of $F_{reattach}$ in the above formula (3.31) as:

$$\gamma_{sep} = s_2 \cdot \min\left[\left(\frac{Re_v}{3.235 \cdot Re_{\theta c}} - 1 \right), 1.0 \right] F_{\theta t} \qquad (3.32a)$$

$$\gamma_{eff} = \max\left(\gamma, \gamma_{sep}\right) \qquad (3.32b)$$

$$s_2 = 1.0 - 2.0 \qquad (3.32c)$$

where s_2 is a constant carefully calibrated for the turbulence models in interest. Because the blending function $F_{reattach}$ is removed from Eq. (3.31), the new separation correction (Eq. 3.32) is able to boost the production of turbulence (turbulent kinetic energy or eddy viscosity) in both laminar and turbulent separated flows.

The effect of the new separation correction method (Eq. 3.32) has been assessed in the calculation of NACA 4412 airfoil (Coles and Wadcock 1979) with a trailing edge flow separation. This is a benchmark case for the turbulence model with validations and verifications posted at the NASA website (http://turbmodels.larc. nasa.gov/naca4412sep_val.html). The chord Reynolds number based on the free-stream velocity is 1.52×10^6 and the Mach number is 0.09 with an angle of attack of $13.87°$. The flow was tripped into turbulent flow at the leading edge in the wind tunnel test, while a turbulent flow separation zone was formed at the trailing edge of the airfoil, as shown in Fig. 3.13.

Computed velocity contours based on the S-A turbulence model are illustrated in Figs. 3.14 and 3.15, and computed Reynolds shear stresses are shown in Figs. 3.16 and 3.17, respectively. These results in Figs. 3.14 and 3.15 clearly show the efficacy of the new separation correction method (Eq. 3.32) in controlling the size of separation zone at the trailing edge of the airfoil. The computation without using the separation correction, shown in Fig. 3.14, shows an overly predicted separation zone compared to the wind tunnel measurement (Fig. 3.13). This is because within the separation zone, the turbulent activity or shear stress is noticeably under-predicted near the wall (Fig. 3.16) by the S-A turbulence model. The computed velocity with the separation correction ($s_2 = 1.25$) shows a reduced separation zone at the trailing edge of the airfoil (Fig. 3.15), because the low-momentum fluids within the separation zone are energized by the increased eddy activities or Reynolds shear stresses (Fig. 3.17).

Computational investigations for the NACA 4412 airfoil as well as other cases have indicated a strong correlation between the level of turbulence (eddy viscosity,

Fig. 3.13 Measured velocity distributions and locations near the trailing edge of NACA 4412 airfoil (*source* http://turbmodels.larc.nasa. gov/naca4412sep_val.html)

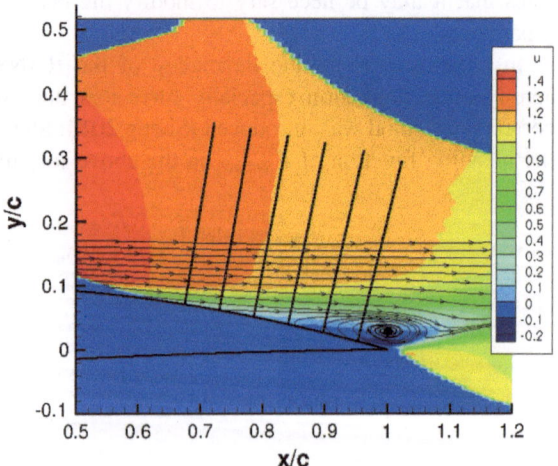

Fig. 3.14 Computed velocity distributions near the trailing edge of NACA 4412 airfoil without separation correction

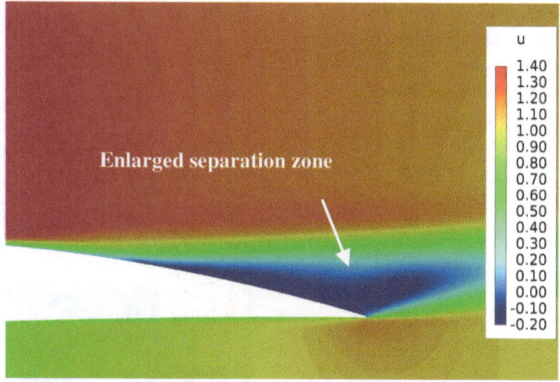

Fig. 3.15 Computed velocity distributions near the trailing edge of NACA 4412 airfoil with separation correction

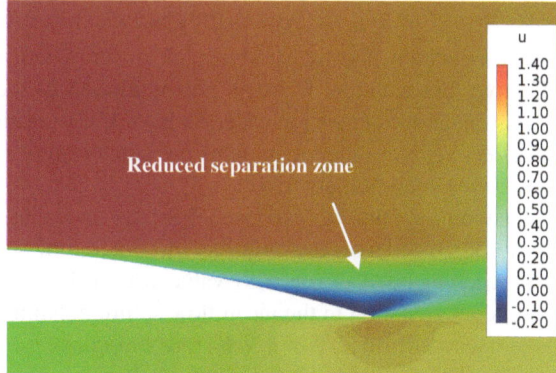

Fig. 3.16 Computed Reynolds shear stress near the trailing edge of NACA 4412 airfoil without separation correction

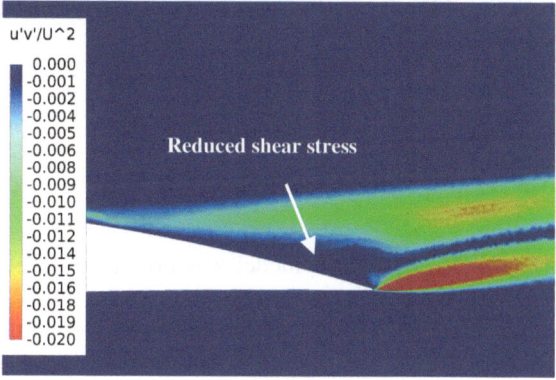

shear stress, or kinetic energy) and the size of the separation zone (Sheng 2014; Sheng et al. 2016). It is worth noting that this new separation correction method can be applied to laminar, transitional, or fully turbulent flows. In the current case of the

Fig. 3.17 Computed
Reynolds shear stress near the
trailing edge of NACA 4412
airfoil with separation
correction

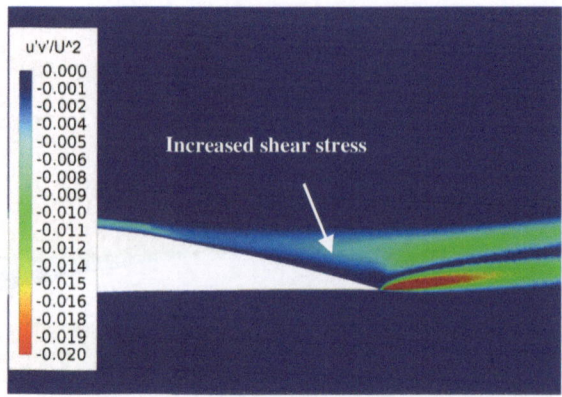

NACA 4412 airfoil, the flow is fully turbulent as it was tripped experimentally. In the computation, the transport equations for γ and $\widetilde{Re}_{\theta t}$ are both solved, but the effective intermittency factor is not multiplied to the turbulence production term unless the above separation criterion Eq. (3.32) is satisfied. Through this treatment, the effective intermittency remains the value of unity except for areas identified by the above separation criterion.

Cautions should be exercised here in choosing the constant value of S_2 in Eq. (3.32). The larger the S_2 value, the smaller the separation bubble or vice versa. It is recommended to use a S_2 value within the specified range in Eq. (3.32c) to avoid contamination to the mean flow solution. Numerical validations of Eq. (3.32) have been performed extensively for both conventional helicopter rotors and pro-protors (Sheng 2014; Sheng et al. 2016), which indicate the necessity of this method to prevent numerically induced premature flow separations for rotors at high thrust levels. For this reason, the above separation correction (Eq. 3.32) is also called the Stall Delay Method (SDM) for the purpose of rotor aerodynamic performance predictions (Sheng 2014; Sheng et al. 2016).

3.3   Integration with the S-A Turbulence Model

The $\gamma - \widetilde{Re}_{\theta t}$ transition model was originally developed by Menter, et al. (2002, 2004, 2006a, b, 2009) under Mether's SST turbulence model. It was later extended to the Spalart-Allmaras (S-A) one-equation turbulence model by Medida and Baeder (2011, 2013) under a structured grid framework, and by Wang and Sheng (2014, 2015) under an unstructured grid framework. In this section, integration with the S-A turbulence model will be presented in order to include transition-modelling capability, whereas the integration with the Menter SST turbulence model will be described in Sect. 3.4.

The Spalart-Allmaras (S-A) turbulence model solves a transport eddy viscosity, referred to as $\tilde{v}$, which relates to the kinematic eddy viscosity (v_t) and the molecular kinematic viscosity (v) as follows:

$$v_t = \tilde{v} f_{v1} \tag{3.33a}$$

$$f_{v1} = \frac{\chi^3}{\chi^3 + c_{v1}^3} \tag{3.33b}$$

$$\chi = \frac{\tilde{v}}{v} \tag{3.33c}$$

where

$$v = \frac{\mu}{\rho}; \quad v_t = \frac{\mu_t}{\rho} \tag{3.33d}$$

The governing equation of the S-A turbulence model is written as:

$$\frac{\partial \tilde{v}}{\partial t} + u_j \frac{\partial \tilde{v}}{\partial x_j} = P_v - D_v + \frac{1}{\sigma}\left[\frac{\partial}{\partial x_j}\left((v + \tilde{v})\frac{\partial \tilde{v}}{\partial x_j}\right) + c_{b2}\frac{\partial \tilde{v}}{\partial x_i}\frac{\partial \tilde{v}}{\partial x_i}\right] \tag{3.34}$$

where P_v and D_v are the production and destruction terms, respectively, which are expressed as:

$$P_v = c_{b1}(f_{r1} - f_{t2})\tilde{S}\tilde{v} + \frac{c_{b1}}{\kappa^2 d^2}\tilde{v}^2 f_{t2} \tag{3.35}$$

$$D_v = [c_{w1}f_w - \frac{c_{b1}}{\kappa^2}f_{t2}](\frac{\tilde{v}}{d})^2 \tag{3.36}$$

$$\tilde{S} = S + \frac{\tilde{v}}{\kappa^2 d^2}f_{v2} \tag{3.37a}$$

$$S = \sqrt{2\Omega_{ij}\Omega_{ij}} \tag{3.37b}$$

$$\Omega_{ij} = \frac{1}{2}(\frac{\partial u_i}{\partial x_j} - \frac{\partial u_j}{\partial x_i}) \tag{3.37c}$$

$$f_{v2} = 1 - \frac{\chi}{1 + \chi f_{v1}} \tag{3.37d}$$

$$f_w = g\left[\frac{1 + c_{w3}^6}{g^6 + c_{w3}^6}\right]^{\frac{1}{6}} \tag{3.37e}$$

$$g = r + c_{w2}(r^6 - r) \tag{3.37f}$$

$$r = \min\left(\frac{\tilde{v}}{\tilde{S}\kappa^2 d^2}, 10\right) \tag{3.37g}$$

$$f_{t2} = c_{t3}\exp(-c_{t4}\chi^2) \tag{3.37h}$$

$$f_{r1} = 1 \tag{3.37i}$$

In the above formulation, d is the distance to the nearest wall surface. The constants in the S-A turbulence model are:

$$\sigma = \frac{2}{3}; C_{b1} = 0.1355; C_{b2} = 0.622$$

$$\kappa = 0.41; C_{w1} = \frac{C_{b1}}{\kappa^2} + \frac{1 + C_{b1}}{\sigma}; C_{w2} = 0.3; C_{w3} = 2.0 \tag{3.38}$$

$$C_{v1} = 7.1; C_{t1} = 1.0; C_{t2} = 2.0; C_{t3} = 1.1; C_{t4} = 2.0;$$

For boundary conditions, Spalart and Rumsey (2007) suggested that the ambient value near the body must be decidedly lower than that in the boundary layer's outer region without affecting the boundary layer through diffusion. Following the recommendations of Spalart and Rumsey (2007), the boundary condition at viscous walls is:

$$\tilde{v}_{wall} = 0; \text{ or } v_{t,wall} = 0 \tag{3.39a}$$

and the far field (or ambient) value should be:

$$\tilde{v}_{farfield} = (3 \sim 5)v_{ref}; \text{ or } v_{t,farfield} = (0.21 \sim 1.29)v_{ref} \tag{3.39b}$$

where v_{ref} is the molecular kinematic viscosity at the reference state.

The production and the destruction terms in the S-A turbulence model are modified by the effective intermittency (γ_{eff}) calculated from Eq. (3.32b), which are written as:

$$\frac{\partial \tilde{v}}{\partial t} + u_j \frac{\partial \tilde{v}}{\partial x_j} = P_{v-TM} - D_{v-TM} + \frac{1}{\sigma}\left[\frac{\partial}{\partial x_j}\left((v + \tilde{v})\frac{\partial \tilde{v}}{\partial x_j}\right) + c_{b2}\frac{\partial \tilde{v}}{\partial x_i}\frac{\partial \tilde{v}}{\partial x_i}\right] \tag{3.40}$$

$$P_{v-TM} = \gamma_{eff} \cdot P_v \tag{3.41}$$

$$D_{v-TM} = \min(\max(\gamma_{eff}, 0.1), 1.0) \cdot D_v \tag{3.42}$$

where P_v and D_v are the production and destruction terms in the original S-A turbulence model.

3.4 Integration with the SST Turbulence Model

Menter (1994) proposed a two-equation Shear Stress Transport (SST) turbulence model that combines the advantages of the k-ε and k-ω turbulence models to achieve an optimal model formulation. To achieve this objective, a blending function (F_1) is introduced that activates the k-ω model in the near wall region and then the k-ε model for the rest of the flow. By this formulation, the robust near-wall performance of the k-ω model is acquired while avoiding the potential error resulting from the free-stream sensitivity of the k-ω model. In addition, the SST model provides a modified definition of the turbulent viscosity to account for the transport effects of the principal turbulent shear stress. This modification is required to accurately capture the onset of separation under pressure gradients. The modeled equations for the turbulent kinetic energy (k) and turbulence frequency (ω) are expressed as:

$$\frac{\partial(\rho k)}{\partial t} + \frac{\partial(\rho U_j k)}{\partial x_j} = P_k - \beta^* \rho \omega k + \frac{\partial}{\partial x_j}\left[(\mu + \sigma_k \mu_t)\frac{\partial k}{\partial x_j}\right] \qquad (3.43)$$

$$\begin{aligned}
\frac{\partial(\rho \omega)}{\partial t} + \frac{\partial(\rho U_j \omega)}{\partial x_j} &= \frac{\gamma}{v_t} P_k - \beta \rho \omega^2 + \frac{\partial}{\partial x_j}\left[(\mu + \sigma_\omega \mu_t)\frac{\partial \omega}{\partial x_j}\right] \\
&\quad + 2(1 - F_1)\frac{\rho \sigma_{\omega 2}}{\omega}\frac{\partial k}{\partial x_j}\frac{\partial \omega}{\partial x_j}
\end{aligned} \qquad (3.44)$$

where P_k is the production term for the turbulent kinetic energy. A production limiter is used to prevent the build-up of turbulence in stagnation regions:

$$\widetilde{P}_k = \mu_t \frac{\partial U_i}{\partial x_j}\left(\frac{\partial U_i}{\partial x_j} + \frac{\partial U_j}{\partial x_i}\right); P_k = \min(\widetilde{P}_k, 10\beta^* \rho \omega k) \qquad (3.45)$$

The turbulent eddy viscosity is defined as:

$$\mu_t = \frac{\rho a_1 k}{\max(a_1 \omega, SF_2)} \qquad (3.46)$$

where S is the invariant of the strain rate, defined as:

$$S = \sqrt{2 S_{ij} S_{ij}} \; ; \quad S_{ij} = \frac{1}{2}\left(\frac{\partial u_i}{\partial x_j} + \frac{\partial u_j}{\partial x_i}\right) \qquad (3.47)$$

and k and ω are the turbulent kinetic energy and frequency, which are related with the turbulence intensity (Tu) and turbulent eddy viscosity (v_t) as:

$$k = \frac{3}{2}(U \cdot Tu)^2; \quad \omega = \frac{k}{v_t} \tag{3.48}$$

In the above equations, F_1 is a blending function, which equals zero away from the wall boundary (k-ε model) and one inside of the boundary layer (k-ω model). F_1 is calculated by the following formula:

$$F_1 = \tanh\{arg_1^4\} \tag{3.49a}$$

$$arg_1 = \min\left[\max\left(\frac{\sqrt{k}}{\beta^* \omega d}, \frac{500v}{d^2\omega}\right), \frac{4\rho\sigma_{\omega2}k}{CD_{k\omega}d^2}\right] \tag{3.49b}$$

$$CD_{kw} = \max\left(2\rho\sigma_{\omega2}\frac{1}{\omega}\frac{\partial k}{\partial x_i}\frac{\partial \omega}{\partial x_i}, 10^{-10}\right) \tag{3.49c}$$

where d is the distance to the nearest wall. F_2 is a second blending function, which is defined as:

$$F_2 = \tanh\{arg_2^2\} \tag{3.50a}$$

$$arg_2 = \max\left(\frac{2\sqrt{k}}{\beta^* \omega d}, \frac{500v}{d^2\omega}\right) \tag{3.50b}$$

All constants are computed by a blend of the corresponding constants of the k-ω model and the k-ε model via $\alpha = \alpha_1 F_1 + \alpha_2(1 - F_1)$, where α_1 and α_2 stand for the coefficients in the two models, respectively. The constants for the SST turbulence model are as follows:

$$\beta^* = 0.09;$$
$$\gamma_1 = \tfrac{5}{9}; \beta_1 = \tfrac{3}{40}; \sigma_{k1} = 0.85; \sigma_{\omega1} = 0.5; \tag{3.51}$$
$$\gamma_2 = 0.44; \beta_2 = 0.0828; \sigma_{k2} = 1; \sigma_{\omega2} = 0.856;$$

The boundary conditions for the turbulent kinetic energy and the turbulence frequency on the wall are given as:

$$k_{wall} = 0; \quad \omega_{wall} = 10\frac{6v}{\beta_1(\Delta d)^2} \tag{3.52}$$

where Δd is the distance between the nearest grid point in the flow field and the wall boundary layer.

Spalart and Rumsey (2007) also made recommendations for the lower and upper bound values at the far field boundary for the SST model. The far field (or ambient) values of k and ω are:

$$\frac{10^{-6}U_{ref}^2}{Re_{ref}} < k_{farfield} < \frac{10^{-4}U_{ref}^2}{Re_{ref}}; \quad \omega_{farfield} = 5\frac{U_{ref}}{L_{ref}}; \quad (3.53a)$$

The equivalent values for the turbulence intensity (Tu) and turbulent eddy viscosity (ν_t) at the far field are:

$$0.063\ \% < Tu_{farfield} < 0.816\ \%$$
$$2 \times 10^{-7}Re_{ref}\nu_{ref} < \nu_{t,farfield} < 2 \times 10^{-5}Re_{ref}\nu_{ref} \quad (3.53b)$$

where Re_{ref} is the reference Reynolds number based on the reference velocity (U_{ref}) and the reference length of the body (L_{ref}), and ν_{ref} is the molecular kinematic viscosity at the reference point.

The $\gamma - \widetilde{Re}_{\theta t}$ transition model has been incorporated into the SST turbulence model by Menter et al. (2002, 2004, 2006a, b). The coupling between the transition model and the turbulence model is accomplished by modifying the production and destruction terms of the turbulent kinetic energy as follows:

$$\frac{\partial(\rho k)}{\partial t} + \frac{\partial(\rho U_j k)}{\partial x_j} = P_{k-TM} - D_{k-TM} + \frac{\partial}{\partial x_j}\left[(\mu + \sigma_k\mu_t)\frac{\partial k}{\partial x_j}\right] \quad (3.54)$$

$$\frac{\partial(\rho\omega)}{\partial t} + \frac{\partial(\rho U_j\omega)}{\partial x_j} = \frac{\gamma}{\upsilon_t}P_k - \beta\rho\omega^2 + \frac{\partial}{\partial x_j}\left[(\mu + \sigma_\omega\mu_t)\frac{\partial\omega}{\partial x_j}\right]$$
$$+ 2(1 - F_1)\frac{\rho\sigma_{\omega2}}{\omega}\frac{\partial k}{\partial x_j}\frac{\partial\omega}{\partial x_j} \quad (3.55)$$

The modified production and destruction terms for the turbulent kinetic energy are calculated by:

$$P_{k-TM} = \gamma_{eff}P_k \quad (3.56)$$

$$D_{k-TM} = \min(\max(\gamma_{eff}, 0.1), 1.0)D_k \quad (3.57)$$

where P_k and D_k are the production and destruction terms from the turbulent kinetic energy equation in the original SST turbulence model. γ_{eff} is the effective intermittency obtained from the transition model. It should be emphasized that the intermittency is used only to control the source terms in the k-equation and is not used to multiply the eddy viscosity.

The final modification to the SST model is a change in the blending function F_1 responsible for switching between the k-ε model and the k-ω model:

$$R_y = \frac{\rho y\sqrt{k}}{\mu} \quad (3.58a)$$

$$F_3 = e^{-(\frac{R_y}{120})^8} \qquad\qquad (3.58b)$$

$$F_1 = \max(F_{1orig}, F_3) \qquad\qquad (3.58c)$$

where F_{1orig} is the original blending function from the SST turbulence model.

3.5   Numerical Procedures

The Langtry and Menter's $\gamma - \widetilde{Re}_{\theta t}$ transition model is solved along with the Spalart-Allmaras (S-A) one-equation turbulence model or with the Menter Shear Stress Transport (SST) two-equation turbulence model in a tightly coupled manner. This forms a three-equation system for the S-A based transition model (SA-TM) or a four-equation system for the SST based transition model (SST-TM). These systems of equations are solved in a unified finite volume method based on an implicit Newton's time marching scheme, similar to the mean flow solver of the Navier-Stokes governing equations (Sheng 2011). In the following sections, the normalized transport equations for the $\gamma - \widetilde{Re}_{\theta t}$ transition model, the S-A turbulence model, and the SST turbulence model are presented first. A unified discretized scheme is then described for solving the system of equations of the three-equation S-A transition model (SA-TM) and the four-equation SST transition model (SST-TM). This includes the flux evaluation on the face of the control volume and the time marching scheme over the computational domain under a general unstructured grid topology.

3.5.1   Normalized Transport Equations

It is helpful to express the transport equations in non-dimensional forms. Here, the following reference quantities are chosen to normalize the variables and the transport equations, where the subscript "*ref*" represents the state of reference values: density (ρ_{ref}); temperature (T_{ref}); pressure $(\rho_{ref} U_{ref}^2)$; velocity (U_{ref}); length (L_{ref}); time (L_{ref}/U_{ref}); eddy viscosity (ν_{ref}); kinetic energy (U_{ref}^2); and turbulence frequency (U_{ref}/L_{ref}).

The non-dimensional form of the S-A turbulence model can be written as:

$$\frac{\partial \widetilde{\nu}}{\partial t} + u_j \frac{\partial \widetilde{\nu}}{\partial x_j} = P_\nu - D_\nu + \frac{1}{\sigma Re_{ref}} \left[\frac{\partial}{\partial x_j} \left((\nu + \widetilde{\nu}) \frac{\partial \widetilde{\nu}}{\partial x_j} \right) + c_{b2} \frac{\partial \widetilde{\nu}}{\partial x_i} \frac{\partial \widetilde{\nu}}{\partial x_i} \right] \qquad (3.59)$$

where the normalized production and destruction terms for the S-A model are:

$$P_v = c_{b1}(f_{r1} - f_{t2})\tilde{S}\tilde{v} + \frac{c_{b1}}{Re_{ref}\kappa^2 d^2}\tilde{v}^2 f_{t2} \tag{3.60}$$

$$D_v = \frac{1}{Re_{ref}}[c_{w1}f_w - \frac{c_{b1}}{\kappa^2}f_{t2}](\frac{\tilde{v}}{d})^2 \tag{3.61}$$

Similarly, the non-dimensional transport equations for the SST model can be written as:

$$\frac{\partial(\rho k)}{\partial t} + \frac{\partial(\rho U_j k)}{\partial x_j} = P_k - \beta^*\rho\omega k + \frac{\partial}{Re_{ref}\partial x_j}\left[(\mu + \sigma_k\mu_t)\frac{\partial k}{\partial x_j}\right] \tag{3.62}$$

$$\frac{\partial(\rho\omega)}{\partial t} + \frac{\partial(\rho U_j\omega)}{\partial x_j} = \frac{\gamma}{v_t}P_k - \beta\rho\omega^2 + \frac{\partial}{Re_{ref}\partial x_j}\left[(\mu + \sigma_\omega\mu_t)\frac{\partial\omega}{\partial x_j}\right]$$
$$+ 2(1 - F_1)\frac{\rho\sigma_{\omega 2}}{\omega}\frac{\partial k}{\partial x_j}\frac{\partial\omega}{\partial x_j} \tag{3.63}$$

where F_1 is the normalized blending function defined by:

$$F_1 = \tanh\left\{\left\{\min\left[\max\left(\frac{\sqrt{k}}{\beta^*\omega y}, \frac{500v}{y^2\omega}Re_{ref}\right), \frac{4\rho\sigma_{w2}k}{CD_{kw}y^2}\right]\right\}\right\} \tag{3.64}$$

The normalized turbulent eddy viscosity is defined as follows:

$$\mu_t = Re_{ref}\frac{a_1 k}{\max(a_1\omega, SF_2)} \tag{3.65}$$

where S is the invariant measure of the strain rate, and F_2 is a second blending function defined by:

$$F_2 = \tanh\left\{[\max\left(\frac{2\sqrt{k}}{\beta^*\omega y}, \frac{500v}{y^2\omega}Re_{ref}\right)]^2\right\} \tag{3.66}$$

The production term for the turbulent kinetic energy is normalized as:

$$\tilde{P}_k = \frac{\mu_t}{Re_{ref}}\frac{\partial U_i}{\partial x_j}\left(\frac{\partial U_i}{\partial x_j} + \frac{\partial U_j}{\partial x_i}\right); P_k = \min(\tilde{P}_k, 10\beta^*\rho\omega k) \tag{3.67}$$

The transport equations for the $\gamma - \widetilde{Re}_{\theta t}$ transition model are also normalized by the above reference values. The normalized transport equations for intermittency (γ) and transition onset momentum thickness Reynolds number can be written as:

$$\frac{\partial(\rho\gamma)}{\partial t} + u_j \frac{\partial(\rho\gamma)}{\partial x_j} = P_\gamma - E_\gamma + \frac{1}{Re_{ref}} \frac{\partial}{\partial x_j} \left[\left(\mu + \frac{\mu_t}{\sigma_\gamma} \right) \frac{\partial\gamma}{\partial x_j} \right] \qquad (3.68)$$

$$\frac{\partial(\rho\widetilde{Re}_{\theta t})}{\partial t} + u_j \frac{\partial(\rho\widetilde{Re}_{\theta t})}{\partial x_j} = P_{\theta t} + \frac{1}{Re_{ref}} \frac{\partial}{\partial x_j} \left[\sigma_{\theta t}(\mu + \mu_t) \frac{\partial}{\partial x_j} \widetilde{Re}_{\theta t} \right] \qquad (3.69)$$

where the normalized production and destruction terms for the intermittency equation are:

$$P_\gamma = F_{length} c_{a1} \rho S [\gamma F_{onset}]^{0.5} (1 - c_{e1}\gamma) \qquad (3.70)$$

$$E_\gamma = c_{a2} \rho \Omega \gamma F_{turb} (c_{e2}\gamma - 1) \qquad (3.71)$$

and the normalized production term for the momentum thickness Reynolds number is:

$$P_{\theta t} = c_{\theta t} \frac{\rho}{t} \left(Re_{\theta t} - \widetilde{Re}_{\theta t} \right) (1 - F_{\theta t}) \qquad (3.72)$$

where F_{length}, F_{onset}, and F_{turb} are the normalized blending functions using the reference values.

The momentum thickness Reynolds number and pressure gradient parameter in the correlation formula (Eqs. 3.19 and 3.20) are normalized based on the following expressions:

$$Re_{\theta t} = Re_{ref} \frac{\rho U \theta}{\mu} \qquad (3.73)$$

$$\lambda_\theta = Re_{ref} \frac{\rho \theta^2}{\mu} \frac{dU}{ds} \qquad (3.74)$$

3.5.2   Discretization Scheme

The normalized transport equations for the three-equation S-A transition model (SA-TM) and four-equation SST transition model (SST-TM) are discretized in a way similar to the mean flow governing equations (Sheng 2011), based on a node-centered finite volume scheme on unstructured grids (Anderson and Bonhaus 1994; Hyams et al. 2000). The nondimensional transport equations can be written in integral form as:

$$\frac{\partial}{\partial t} \iiint q \, dv + \oiint \left[\left(\vec{F}_c \cdot \vec{n} \right) dA + \left(\vec{F}_d \cdot \vec{n} \right) dA \right] = \iiint S \, dv \qquad (3.75)$$

where q is a vector of the primitive variables. For the three-equation S-A transition model:

$$q = \begin{bmatrix} \tilde{v} \\ \gamma \\ \widetilde{Re}_{\theta t} \end{bmatrix} \qquad (3.76a)$$

and for the four-equation SST $k - \omega$ transition model:

$$q = \begin{bmatrix} k \\ \omega \\ \gamma \\ \widetilde{Re}_{\theta t} \end{bmatrix} \qquad (3.76b)$$

$\vec{F}_c$ and $\vec{F}_d$ are the vectors of convective and diffusive fluxes evaluated on the face of a control volume, respectively. The right hand side of Eq. (3.75), S, is a vector of the source terms, which combines both production and destruction terms of the transport equations. It should be noted that all quantities in Eqs. (3.75) and (3.76) are normalized as the non-dimensional variables.

The general discretized system of equations at a finite volume (i), including the integration of the convective and diffusive fluxes on the control surface (j), and the source term at the control volume (i), can be expressed as:

$$\rho_i \frac{\Delta q_i}{\Delta t} \Delta V_i + \sum_{j=1}^{n_j} \left[\left(\vec{F}_c \cdot \vec{n} \right)_j \Delta A_j + \left(\vec{F}_d \cdot \vec{n} \right)_j \Delta A_j \right] = S_i \Delta V_i \qquad (3.77)$$

The subscript i denotes the vertex of the control volume, whose volume is ΔV_i. The index $j = 1, \ldots, n_j$ denotes the jth dual-face ΔA_j with the unit normal vector $\vec{n}$. The total number of dual faces for the control volume (i) is nj.

3.5.3 Flux Evaluation

The convective fluxes of the transport equations are calculated using a formula similar to the Roe's flux approximation (Roe 1981). Consider the control surface (j) at the control volume (i), whose left and right states are denoted by L and R. The numerical fluxes projected on the control surface (j) can then be written as:

$$\vec{F_c} \cdot \vec{n} = \frac{1}{2}(F(q_L) + F(q_R)) - \frac{1}{2}\rho_{avg} \left| \Theta_{avg} \right| (q_R - q_L) \qquad (3.78)$$

where $F(q_L)$ and $F(q_R)$ are the convective fluxes evaluated at the left and right nodes of the control surface (j), which have two expressions for the S-A and SST based transition models, respectively. For the S-A based transition model:

$$F(q) = \begin{bmatrix} \rho \tilde{v} \Theta \\ \rho \underline{\gamma} \Theta \\ \rho \widetilde{Re}_{\theta t} \Theta \end{bmatrix} \qquad (3.79)$$

and for the SST based transition model:

$$F(q) = \begin{bmatrix} \rho k \Theta \\ \rho \omega \Theta \\ \rho \underline{\gamma} \Theta \\ \rho \widetilde{Re}_{\theta t} \Theta \end{bmatrix} \qquad (3.80)$$

where Θ is a contravariant velocity at the control surface (j). Both ρ_{avg} and Θ_{avg} are evaluated using the averaged densities and contravariant velocities at the left and right states. All variables in the flux formulas (3.79) and (3.80) are non-dimensional quantities normalized by the reference values introduced at the beginning of this section.

For the diffusive fluxes ($\vec{F_d} \cdot \vec{n}$) in the discretized equation (3.77), a directional derivative method proposed by Hyams et al. (2000) is employed for calculating the diffusive fluxes on general mixed element unstructured grids:

$$\vec{F_d} \cdot \vec{n} = \frac{1}{Re_{ref}} \left[\overline{\nabla q} + \left(q_R - q_L - \overline{\nabla q} \cdot \vec{s} \right) \right] \frac{\vec{s}}{\left| \Delta \vec{s} \right|^2} \qquad (3.81)$$

where $\vec{s}$ is a unit vector in the direction of the edge (j), $\left| \Delta \vec{s} \right| = \left| \vec{x}_R - \vec{x}_L \right|$ which is the length of the edge, and $\overline{\nabla q} = (\nabla q_R + \nabla q_L)/2$, where ∇q is the gradient of the transport quantities expressed in Eq. (3.76).

3.5.4 Time Marching Method

The implicit time marching scheme is used to solve the discretized system of equations (Eq. 3.77). The temporal discretization of the transport equations can be written as:

$$\rho_i \frac{\Delta q_i^n - \frac{\theta}{1-\theta}\Delta q_i^{n-1}}{\Delta t} + \frac{1}{\Delta V_i}\sum_{j=1}^{n_j}[\left(\vec{F}\cdot\vec{n}\right)_j^{n+1}\Delta A_j] = S_i^{n+1} \qquad (3.82)$$

where $\Delta q_i^n = q_i^{n+1} - q_i^n$ and $\Delta q_i^{n-1} = q_i^n - q_i^{n-1}$. Δt is the time step increment between the time steps n and $n+1$. $\vec{F}\cdot\vec{n}$ are the combination of both convective and diffusive fluxes of the discretized transport equations. The constant θ is used to control the order of temporal accuracy. A first order temporal accuracy of the Euler implicit scheme is given by the choice of $\theta = 0$. Correspondingly, a second-order time accurate Euler implicit scheme is given by $\theta = 1$.

The discretized equation (3.82) is solved by Newton's method. The function $N\left(q_i^{n+1}\right)$ is defined as:

$$N\left(q_i^{n+1}\right) = \rho_i \frac{\Delta q_i^n - \frac{\theta}{1-\theta}\Delta q_i^{n-1}}{\Delta t} + \frac{1}{\Delta V_i}\sum_{j=1}^{n_j}[\left(\vec{F}\cdot\vec{n}\right)_j^{n+1}\Delta A_j] - S_i^{n+1} \qquad (3.83)$$

The Newton's method for this equation can be written as:

$$N'\left(q_i^{n+1,m}\right)\left(q_i^{n+1,m+1} - q_i^{n+1,m}\right) = -N\left(q_i^{n+1,m}\right) \qquad (3.84a)$$

$$q_i^{n+1,m+1} = q_i^{n+1,m} - \frac{N\left(q_i^{n+1,m}\right)}{N'\left(q_i^{n+1,m}\right)} \qquad (3.84b)$$

where $m = 1, 2, 3, \ldots$ are the Newton sub-iteration steps, with an initial guess of $q_i^{n+1,0} = q_i^n$. The Jacobian matrix of the system of equations can be written as:

$$N'\left(q_i^{n+1}\right) = \rho_i \frac{I}{\Delta t} + \frac{1}{\Delta V_i}\sum_{j=1}^{n_j}\frac{\partial\left(\vec{F}\cdot\vec{n}\right)_j^{n+1}}{\partial q_i^{n+1}}\Delta A_j - \frac{\partial S_i^{n+1}}{\partial q_i^{n+1}} \qquad (3.85)$$

where the first term on the right-hand side is the contribution from the unsteady time derivative of q, the second term is the contribution from the steady state residual of the transport equations (including both convective and diffusive terms), and the last term is the contribution from the source term. The inviscid flux Jacobian is evaluated by taking the derivative of convective fluxes with respect to q, which can be written as:

$$\sum_{j=1}^{n_j}\frac{\partial\left(\vec{F}_c\cdot\vec{n}\right)_j^{n+1}}{\partial q^{n+1}}\Delta A_j = \sum_{j=1}^{n_j}\frac{\partial\left(\vec{F}_c\cdot\vec{n}\right)_j^{n+1}}{\partial q_L^{n+1}}\Delta A_j + \sum_{j=1}^{n_j}\frac{\partial\left(\vec{F}_c\cdot\vec{n}\right)_j^{n+1}}{\partial q_R^{n+1}}\Delta A_j \qquad (3.86a)$$

$$\frac{\partial \left(\vec{F}_c \cdot \vec{n}\right)_j^{n+1}}{\partial q_L^{n+1}} = \frac{1}{2}\left(\frac{\partial \left(\vec{F}(q_L) \cdot \vec{n}\right)^{n+1}}{\partial q_L^{n+1}} + \rho_{avg}\left|\Theta_{avg}\right|\right) \tag{3.86b}$$

$$\frac{\partial \left(\vec{F}_c \cdot \vec{n}\right)_j^{n+1}}{\partial q_R^{n+1}} = \frac{1}{2}\left(\frac{\partial \left(\vec{F}(q_R) \cdot \vec{n}\right)^{n+1}}{\partial q_R^{n+1}} - \rho_{avg}\left|\Theta_{avg}\right|\right) \tag{3.86c}$$

Similarly, the diffusive flux Jacobian is evaluated by taking the derivative of the diffusive fluxes with respect to q. The Jacobian of the source term is evaluated in a similar way in order to enhance the stability of the numerical scheme. It is recommended that only the positive contributions to the source-term Jacobian matrix be considered. This is done by ignoring the terms that are not guaranteed to be positive. This procedure leads to an increase in the diagonal dominance of the system matrix and thereby enhances its stability.

The nonlinear system of Eq. (3.83) is solved over the computational domain, which results in a sparse system of equations at each time. The solution of the sparse system of Eq. (3.83) is obtained by a relaxation scheme, where $\Delta q_i^{n+1,m}$ is obtained through a sequence of iterations, $\left\{\Delta q_i^{n+1,m}\right\}^i$, which converge to $\Delta q_i^{n+1,m}$. There are several variations of classic relaxation procedures for solving this linear system of equations. However, a symmetric implicit Gauss-Seidel procedure is suggested. To clarify the scheme, the system matrix is first written as a linear combination of matrices representing the diagonal, upper triangular and lower triangular parts at each time step:

$$[A] = [D + U + L] \tag{3.87a}$$

where each part is defined as:

$$[D] = \rho_i \frac{I}{\Delta t} + \frac{1}{\Delta V_i}\sum_{j \in N(0)} \frac{\partial \left(\vec{F} \cdot \vec{n}\right)_j^{n+1}}{\partial \mathbf{q}^{n+1}} \Delta A_j - \frac{\partial S_i^{n+1}}{\partial q^{n+1}} \tag{3.88b}$$

$$[U] = \frac{1}{\Delta V_i}\sum_{j \in N_U(0)} \frac{\partial \left(\vec{F} \cdot \vec{n}\right)_j^{n+1}}{\partial q^{n+1}} \Delta A_j \tag{3.88c}$$

$$[L] = \frac{1}{\Delta V_i}\sum_{j \in N_L(0)} \frac{\partial \left(\vec{F} \cdot \vec{n}\right)_j^{n+1}}{\partial q^{n+1}} \Delta A_j \tag{3.88d}$$

By letting $\{R^{n+1,m}\}$ be the vector of unsteady residuals and $\{\Delta q^{n+1,m}\}$ represent the change in the dependent variables, the symmetric Gauss-Seidel relaxation can be written as the following two-step procedures:

$$[L+D]\{\Delta q^{n+1,m}\}^{i+1/2} + [U]\{\Delta q^{n+1,m}\}^{i} = \{R^{n+1,m}\} \qquad (3.89a)$$

$$[D+U]\{\Delta q^{n+1,m}\}^{i+1} + [L]\{\Delta q^{n+1,m}\}^{i+1/2} = \{R^{n+1,m}\} \qquad (3.89b)$$

In the forward pass, the $\{\Delta q^{n+1,m}\}^{i+1/2}$ terms are obtained with the previously updated $\{\Delta q^{n+1,m}\}^{i}$ terms, which were set to zero at the initial stage. In the backward pass, $\{\Delta q^{n+1,m}\}^{i}$ terms are obtained with the most recent value of $\{\Delta q^{n+1,m}\}^{i+1/2}$ from the previous forward pass. Normally eight to twelve symmetric Gauss-Seidel sub-iterations are adequate at each time step to converge the transport equations for either the S-A or the SST-based transition model.

References

Abu-Ghannam BJ, Shaw R (1980) Natural transition of boundary layers—the effects of turbulence, pressure gradient, and flow history. J Mech Eng Sci 22(5):213–228

Anderson WK, Bonhaus DL (1994) An implicit upwind algorithm for computing turbulent flows on unstructured grids. Comput Fluids 23:1–21

Aupoix B, Arnal D, Bezard H et al (2011) Transition and turbulence modeling. J Aerosp Lab, issue 2, Mar 2011

Coles D, Wadcock AJ (1979) Flying-hot-wire study of flow past an NACA 4412 airfoil at maximum lift. AIAA Journal, 17(4):321–329

Driest ER, Blumer CB (1963) Boundary layer transition: freestream turbulence and pressure gradient effects. AIAA J 1(6):1303–1306

Gardarein P, Le Pape A (2016) Numerical simulation of hovering S-76 helicopter rotor including far-field analysis. AIAA-2016-00034. Presented at 54th AIAA Aerospace Sciences Meeting AIAA SciTech Conferences, Sab Diego, California, 4–8 Jan 2016

Hyams D, Sreenivas K, Sheng C et al (2000) An investigation of parallel implicit solution algorithms for incompressible flows on unstructured topologies. AIAA-2000-0271, the 38th AIAA Aerospace Sciences Meeting, Reno, Nevada, 10–13 Jan 2000

Jung MK, Hwang JY, Kwon OJ (2014) Assessment of rotor aerodynamic performances in hover using an unstructured mixed mesh method. Paper presented at the 52nd AIAA Aerospace Sciences Meeting, AIAA SciTech, 13–17 Jan 2014

Langtry RB, Sjolander SA (2002) Prediction of transition for attached and separated shear layers in turbomachinery. AIAA-2002-3643. 38th AIAA/ASME/SAE/ASEE Joint Propulsion Conference and Exhibit

Langtry RB (2006) A correlation-based transition model using local variables for unstructured parallelized CFD codes. PhD Dissertation, University of Stuttgart. http://elib.uni-stuttgart.de/opus/volltexte/2006/2801/

Langtry RB, Menter FR (2006) Predicting 2D airfoil and 3D wind turbine rotor performance using a transition model for general CFD codes. Paper presented at 44th AIAA aerospace sciences meeting and exhibit, Reno, Nevada

Langtry RB, Menter FR (2009) Correlation-based transition modeling for unstructured parallelized computational fluid dynamics codes. AIAA J 47(12):2894–2906

Langtry RB, Menter FR, Likki SR et al (2006) A correlation based transition model using local variables part 2—test cases and industrial applications. ASME J Turbomach 128(3):423–434

Medida S, Baeder JD (2011) Application of the correlation-based γ-$(\overline{Re}_{\theta t})$ transition model to the Spalart-Allmaras turbulence model. Presented at 20th AIAA computational fluid dynamics conference, Honolulu, Hawaii, June 2011

Medida S, Baeder JD (2013) Role of improved turbulence and transition modeling methods in rotorcraft simulations. In: Proceedings of the AHS international 69th annual forum, Phoenix, Arizona, May 2013

Menter FR (1994) Two-equation eddy-viscosity turbulence models for engineering application. AIAA J 32(8):1598–1605

Menter FR, Esch T, Kubachi S (2002) Transition modeling based on local variables. Paper presented at 5th international symposium on engineering turbulence modeling and measurements, Mallorca, Spain

Menter FR, Langtry RB, Likki SR et al (2004) A correlation based transition model using local variables part 1—model formulation. ASME-GT2004-53452. In: Proceedings of ASME TURBO EXPO 2004, Vienna, Austria

Menter FR, Langtry RB, Volker S (2006a) Transition modelling for general purpose CFD codes. J Flow Turbul Combust 77(1–4):277–303

Menter FR, Langtry RB, Likki SR et al (2006b) A correlation based transition model using local variables part 1—model formulation. ASME J Turbomach 128(3):413–422

Menter FR, Smirnov PE, Liu T et al (2015) A one-equation local correlation-based transition model. In: Flow Turbulence and Combustion, 05 July 2015, doi:10.1007/s10494-015-9622-4

Min BY, Wake B (2016) Parametric validation study for a hovering rotor using UT-GENCAS. AIAA-2016-0301, AIAA 2016 SciTech Conferences, Sab Diego, California, 4–8 Jan 2016

Roe PL (1981) Approximate Riemann solvers, parameter vectors and difference schemes. J Comput Phys 43:357–372

Sheng C (2011) A preconditioned method for rotating flows at arbitrary Mach number. Model Simul Eng 537464. doi:10.1155/2011/537464

Sheng C (2014) Predictions of JVX rotor performance in hover and airplane mode using high-fidelity unstructured grid CFD solver. In: Proceedings of the AHS 70th annual forum, Montreal, Canada, 20–22 May 2014

Sheng C, Wang J, Zhao Q (2016) Improved rotor hover predictions using advanced turbulence modeling. J Aircraft. doi:10.2514/1.C033512

Spalart PR, Allmaras SR (1994) A one-equation turbulence model for aerodynamic flows. La Recherche Aerospatiale 1994(1):5–21

Spalart PR, Rumsey CL (2007) Effective inflow conditions for turbulence models in aerodynamic calculations. AIAA J 45(10):2533–2544

Suluksna K, Dechaumphai P, Juntasaro E (2009) Correlation for modeling transitional boundary layers under influences of free stream turbulence and pressure gradient. Int J Heat Fluid Flow 30(1):66–75

Wang J, Sheng C (2014) Validations of a local correlation-based transition model using an unstructured grid CFD solver. Paper presented at AIAA 2014 aviation conferences, Atlanta, Georgia, 16–20 June 2014

Wang J, Sheng C (2015) A comparison of a local correlation-based transition model coupled with SA and SST turbulence models. Paper presented at AIAA 2015 SciTech conferences, Kissimmee, Florida, 5–9 Jan 2015

White FM (2006) Viscous fluid flow, 3rd edn. McGraw-Hill Companies, Inc., New York, pp 231–233

Chapter 4
Validations in 2-D Flows

Abstract Transition models are validated using the benchmark flat plate cases and two-dimensional airfoils in this chapter. Validations are performed for the European Research Community on Flow, Turbulence and Combustion T3 series of the experimental flat plat cases, the Schubauer and Klebanoff flat plate, as well as several two-dimensional airfoils including Aerospatiale-A, VA-2, S809, and NACA4412. The effect of the Stall Delay Method is demonstrated on the NACA4412 airfoil with the trailing edge flow separation.

4.1 Description

In Chap. 3, the Langtry and Menter $\gamma - \widetilde{Re}_{\theta t}$ transition model (Langtry and Menter 2009) is integrated into two widely used turbulence models used in the CFD community today: the Spalart-Allmaras (S-A) one-equation turbulence model (Spalart and Allmaras 1994) and Menter's Shear Stress Transport (SST) two-equation turbulence model (Menter 1994). The performance of the S-A and SST based transition models will be discussed in this chapter. The focus here is to demonstrate the models' efficacy to capture the boundary layer transition phenomena under various flow conditions, such as different free-stream turbulence intensities, pressure gradients, Reynolds number effects, and flow angles of incidence, etc. Comparative studies are performed to investigate the behavior and merit of the two transition modeling approaches under the same CFD framework.[1]

---

[1]Original work was presented at the 53rd American Institute of Aeronautics and Astronautics (AIAA) Aerospace Sciences Meeting, 5–9 January 2015, Kissimmee, Florida, U.S.A.

© The Author(s) 2017   55
C. Sheng, *Advances in Transitional Flow Modeling*, SpringerBriefs
in Applied Sciences and Technology, DOI 10.1007/978-3-319-32576-7_4

4.2 Flat Plates

The European Research Community on Flow, Turbulence and Combustion (ERCOFTAC) T3 series of experimental tests on flat plates (Savill 1993a, b) are the benchmark cases that have been widely used for validating the transition and turbulence models in the CFD community. The experimental investigations of the T3 series were carried out by researchers at Rolls-Royce in the early 1990s, where a flat plate with a length of 1.65 m was mounted to the test section in the wind tunnel under various mean flow and turbulence conditions. The T3 series of flat plates include T3A, T3B, T3AM, T3C2, T3C3, T3C4 and T3C5. Cases of T3A, T3B and T3AM were measured at a zero pressure gradient and each under a different free-stream turbulence intensity of 3, 6.5 and almost 1 %, respectively. Cases on the T3C series combined influences of the free-stream turbulence intensities and pressure gradients, which represents the flow conditions of the aft-loaded turbine blades (Suzen et al. 2002). The major difference among the T3C series is the free-stream velocities and the Reynolds numbers, where the free-stream turbulence intensity is maintained at nearly the same level of 3 %. In addition, the Schubauer and Klebanoff (S&K) flat plate experiment is chosen here in order to test the model's capability to predict natural transition under a free-stream turbulence intensity of less than 1 % (Schubauer and Klebanoff 1956; Dhawan and Narasimha 1958). Table 4.1 provides a summary of the inlet flow conditions for all test cases on the flat plate.

Three unstructured computational meshes are generated for the flat plate cases. The first one is meshed for the zero-pressure gradient cases of T3A, T3B, T3AM and S&K, which is comprised of a flat plate with a length of 1.65 m and a symmetric flat top surface at a height of 0.25 m. The second mesh is generated for the flat plate with different pressure gradients for T3C2, T3C3 and T3C5, which consists of a flat plate wall with a length of 1.65 m and a slip top surface with a profile to match the pressure distribution in the experiment (Suluksna et al. 2009). The third mesh is generated for the T3C4 case only, which is similar to the second

Table 4.1 Inlet conditions of flat plate test cases

Test case	Inlet velocity (m/s)	Inlet turbulence intensity (%)	μ_t/μ	Density (kg/m^3)	Molecular viscosity (kg/ms)
T3A	5.4	3.3	12.0	1.2	1.8×10^{-5}
T3B	9.4	6.5	100.0	1.2	1.8×10^{-5}
T3AM	19.8	0.874	8.72	1.2	1.8×10^{-5}
S&K	50.1	0.3	1.0	1.2	1.8×10^{-5}
T3C2	5.29	3.0	11.0	1.2	1.8×10^{-5}
T3C3	4.0	3.0	6.0	1.2	1.8×10^{-5}
T3C4	1.37	3.0	0.009	1.2	1.8×10^{-5}
T3C5	9.0	4.0	15.0	1.2	1.8×10^{-5}

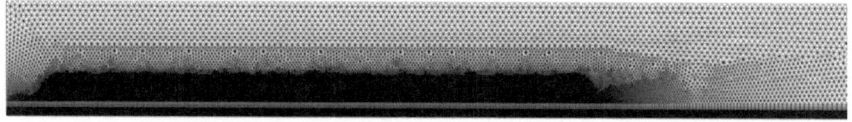

(a) Computational domain for zero-pressure gradient flat plate

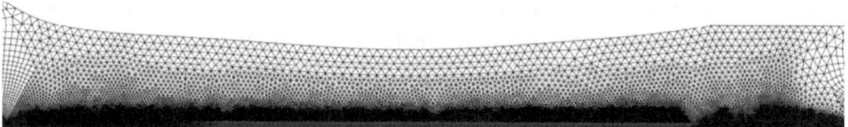

(b) Computational domain for non-zero pressure gradient flat plate

Fig. 4.1 The computational domains and grids for flat plate

mesh but has a different top surface profile. The computational domain has an inlet surface located 0.04 m upstream of the leading edge of the flat plate in order to match the inlet conditions (such as the turbulence intensity) corresponding to the experiment. To better capture the viscous effects in the boundary layer, an anisotropic grid topology is used to generate the boundary layer mesh, with an expansion ratio of 1.1 to ensure sufficient grid resolutions in the normal direction of the wall. The first boundary layer node location is set to ensure that the y^+ value is less than one based on the free-stream Reynolds numbers. The CFD meshes for both zero and non-zero pressure gradient cases are shown in Fig. 4.1.

To reproduce the required pressure gradients from the experimental data of the T3C series, the height of the cross-section along the flow direction was obtained by Suluksna et al. (2009) based on the continuity equation. Two profiles for the computational domains, one representing the T3C4 case and the other representing the rest of the T3C series, are expressed in Eqs. (4.1a) and (4.1b), respectively. Note that in the expressions, h is the height of the cross-section and D is the inlet height ($D = 0.3$ m). It shows that the domain cross-section initially converges and then diverges subsequently to produce a varying pressure along the stream-wise direction. The pressure gradient is negative for all locations where x is less than 0.9 m, and becomes positive afterwards. The upper surface profile for the T3C4 computational domain is defined as:

$$\frac{h}{D} = \min(1.356x^6 - 7.591x^5 + 16.513x^4 - 17.510x^3 + 9.486x^2$$
$$- 2.657x + 0.991, 1.0) \tag{4.1a}$$

and the upper surface profile for the rest of the T3C series is:

$$\frac{h}{D} = \min(1.231x^6 - 6.705x^5 + 14.061x^4 - 14.113x^3 + 7.109x^2$$
$$- 1.900x + 0.950, 1.0) \tag{4.1b}$$

The boundary conditions are specified as follows: A constant far field velocity is set at the inlet of the computational domain. In the rectangular computational domain (zero-pressure gradient cases), a symmetric condition is set for the upper surface. In the varying cross-section computational domain (non-zero pressure gradient cases), a slip wall is set for the curved upper surface. A slip condition is also set to fore and aft of the no-slip flat plate. At the outlet of the computational domain, a far field condition is specified. Figure 4.2 shows computed and measured local free-stream velocities along the flat plate stream-wise direction for the T3C series, where symbols denote the experimental data, and two solid lines denote the computed free-stream velocities. There are some discrepancies between the computed and measured velocity distributions along the flat plate, indicating that the cross-section profiles provided by Eqs. (4.1a) and (4.1b) do not generate the exact pressure gradient distributions as measured in the wind tunnel. This may cause some numerical errors in the comparison of the CFD results with the experimental data for the T3C series below.

The skin friction coefficient C_f is an important parameter that indicates whether the state of the boundary layer is in a laminar, turbulent, or transitional flow regime. The skin friction distribution is plotted as a function of the local Reynolds number, defined as:

$$Re_x = \frac{\rho_\infty Ux}{\mu_\infty} \tag{4.2}$$

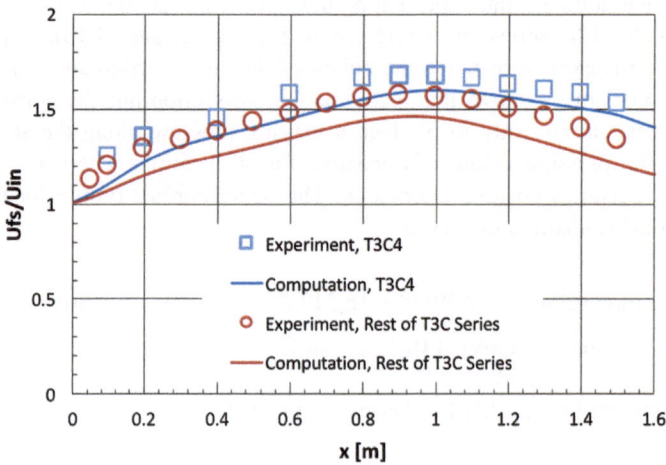

Fig. 4.2 Comparison of computed and measured free-stream velocity distributions for the T3C series

where ρ_∞ and μ_∞ are the inlet free-stream density and dynamic viscosity, respectively, U is the local free-stream velocity, and x is the local stream-wise location to the leading edge of the flat plate.

Both the S-A and SST based $\gamma - \widetilde{Re}_{\theta t}$ transition models described in Chap. 3 are assessed for their ability to predict the boundary layer transition on all the flat plate cases. The inlet free-stream conditions, including the mean velocity, turbulence intensity and turbulent eddy viscosity, are given in Table 4.1. A key difference between the SA-TM and SST-TM transition models is the calculation of the local free-stream turbulence intensity, which is calculated from the kinetic energy term in the SST transition model but is estimated locally based on Eq. (3.30) in the S-A transition model. Computations are assessed between the two models in capturing the onset and length of the transitions, which are the two most important parameters in boundary layer transition phenomena.

4.2.1 Zero-Pressure Gradients

The S&K case (Schubauer and Klebanoff 1956) and ERCOFTAC T3A, T3B, T3AM cases (Savill 1993a, b) are computed at a zero-pressure gradient, where different velocities and free-stream turbulence intensities are specified at the inlet (Table 4.1). Comparisons of the skin friction coefficients between the wind tunnel tests and the computations by both S-A and SST transition models are shown in Figs. 4.3, 4.4, 4.5 and 4.6. For the S&K case ($Tu_\infty = 0.3\,\%$), which was measured in a very quiet wind tunnel with less than 1 % of the turbulence intensity, the natural transition in the boundary layer occurs around $Re_x = 3 \times 10^6$. Both S-A and SST transition models are able to accurately predict the transition onset location, as shown in Fig. 4.3. Because of the very low free-stream turbulence intensity, the

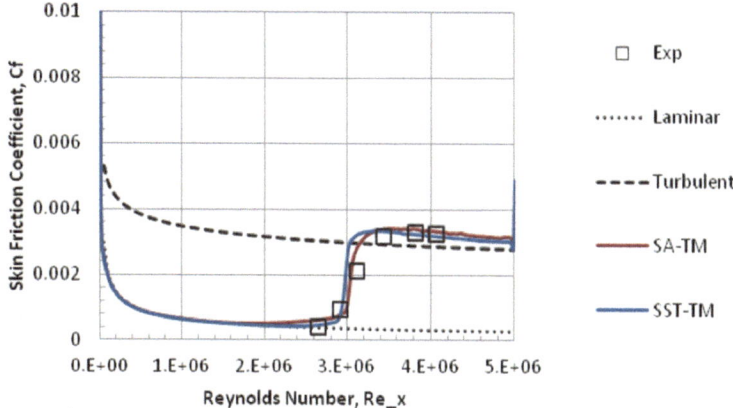

Fig. 4.3 Comparison of skin friction coefficient profiles for S&K case on flat plate

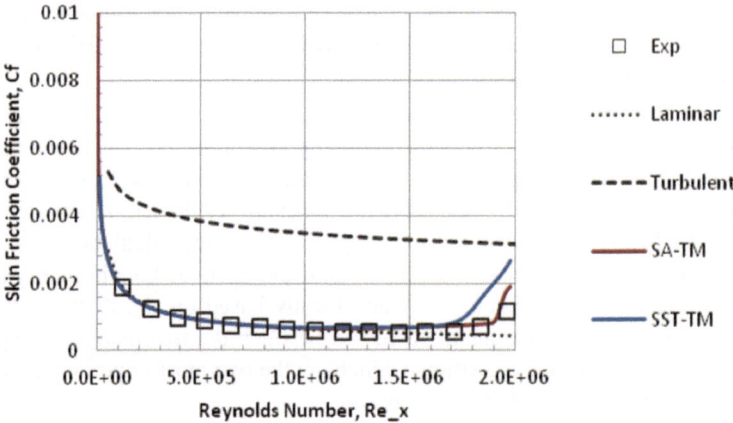

Fig. 4.4 Comparison of skin friction coefficient profiles for T3AM case on flat plate

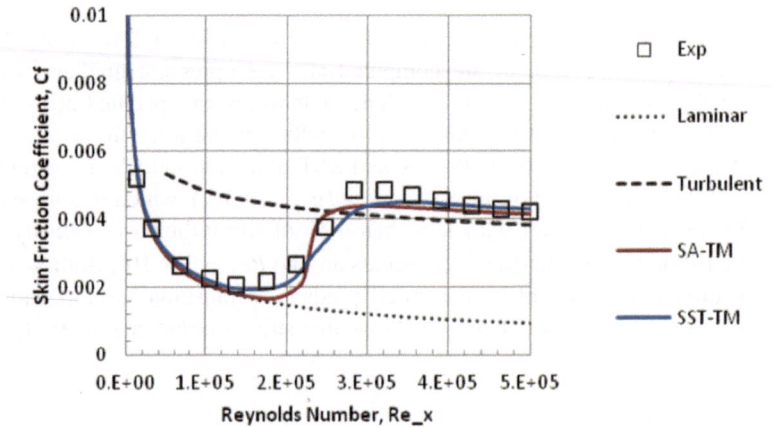

Fig. 4.5 Comparison of skin friction coefficient profiles for T3A case on flat plate

S-A and SST transition models have demonstrated nearly the same behavior in capturing the onset and the growth rate of this natural transition. T3AM ($Tu_\infty = 0.874\%$) is also considered as a low free-stream turbulence intensity case, but has a lower inlet velocity. As such, the transition onset is delayed to downstream of the flat plate compared to the S&K case. The predicted transition onsets by both the S-A and SST transition models match reasonably well with the measurement, as shown in Fig. 4.4. Moving to the moderate free-stream turbulence intensity case, T3A ($Tu_\infty = 3.3\%$) shows the typical bypass transition starting at a Reynolds number of 2×10^5 and ending at 3×10^5 in Fig. 4.5. Both S-A and SST transition models are capable of predicting the correct transition onset locations, but a slightly faster growth rate in the transition region is predicted by the S-A based transition

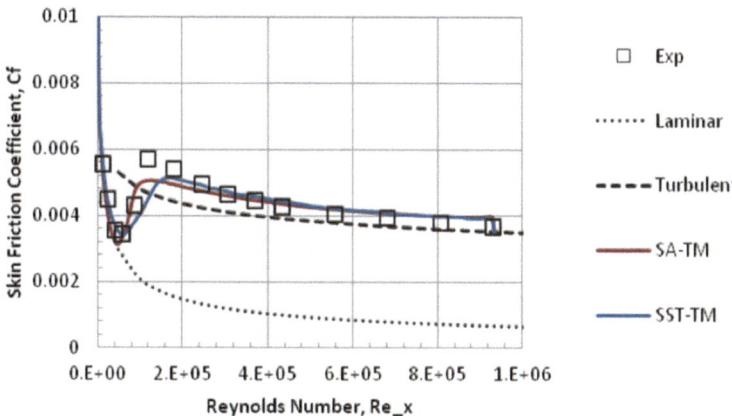

Fig. 4.6 Comparison of skin friction coefficient profiles for T3B case on flat plate

model than by the SST-based model, due to different treatments of the local free-stream turbulence intensity. At the highest free-stream turbulence intensity case, shown in Fig. 4.6, T3B ($Tu_\infty = 6.3\%$) has an early transition close to the leading edge of the flat plate. The S-A and SST based transition models also capture the same trend as observed in the experiment, although a faster growth of turbulence is still observed using the S-A transition model. A major conclusion can be obtained from the zero-pressure gradient flat plate validation cases, which is that both the S-A and SST transition models are capable of predicting the natural and bypass transition phenomena at different levels of free-stream turbulence intensity. However, the S-A based transition model generally predicts a faster turbulence growth rate or a shorter transition region than the SST based transition model. Overall, the SST-based transition model provides a better correlation with the experimental data for the flat plate cases at the zero-pressure gradient.

4.2.2 Non-zero Pressure Gradients

Computations for the T3C series with non-zero pressure gradients are shown in Figs. 4.7, 4.8, 4.9 and 4.10. These test cases are used to validate the transition models in capturing the transition at a moderate free-stream turbulence intensity under the influence of pressure gradients. The case of T3C2 ($Tu_\infty = 3.0\%$) represents a bypass transition that occurs almost at the suction peak (0.9 m from the leading edge of the plate), as shown in Fig. 4.7. Both S-A and SST transition models have correctly predicted the transition onset location. The case for T3C3 ($Tu_\infty = 3.0\%$) has a lower Reynolds number than the T3C2 case due to a reduced inlet velocity. As seen in Fig. 4.8, the transition onset moves toward downstream and occurs in the adverse pressure gradient region. Both the S-A and SST transition

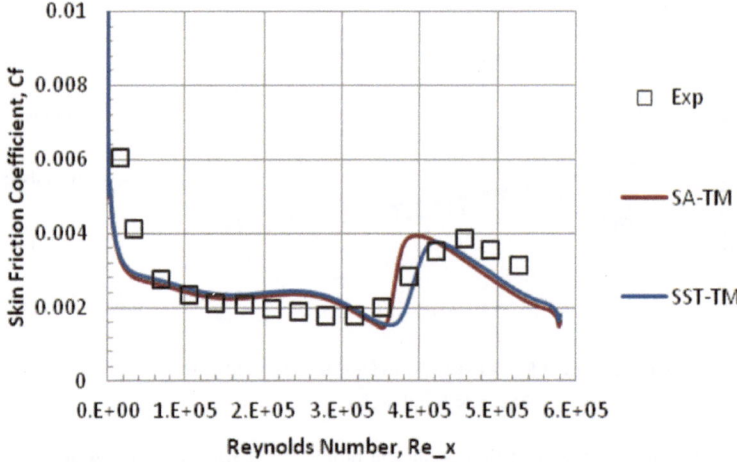

Fig. 4.7 Comparison of skin friction coefficient profiles for T3C2 case on flat plate

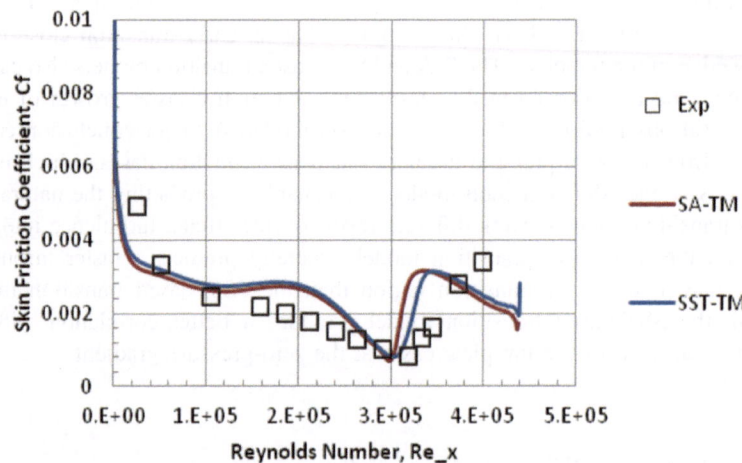

Fig. 4.8 Comparison of skin friction coefficient profiles for T3C3 case on flat plate

models have predicted nearly the same transition onset locations as the pressure gradient increases, but a slightly faster growth rate of turbulence than that observed in the experiment. The comparison of the skin friction coefficient profiles for the T3C5 case ($Tu_\infty = 4.0\%$) is shown in Fig. 4.9. Because it has the highest free-stream velocity in the T3C series, the transition occurs near the leading edge where the flow has a favorable pressure gradient. Both the S-A and SST transition models have correctly responded to the increased Reynolds number and have predicted the transition onset reasonably well. However, the S-A transition model seems to predict a slightly earlier transition onset than the SST transition model.

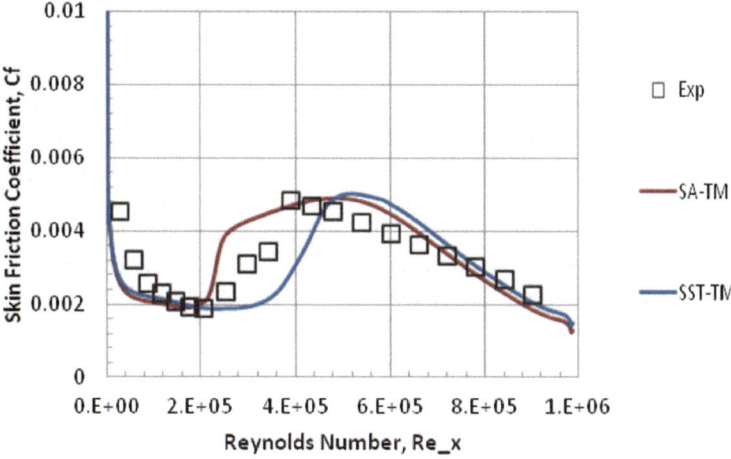

Fig. 4.9 Comparison of skin friction coefficient profiles for T3C5 case on flat plate

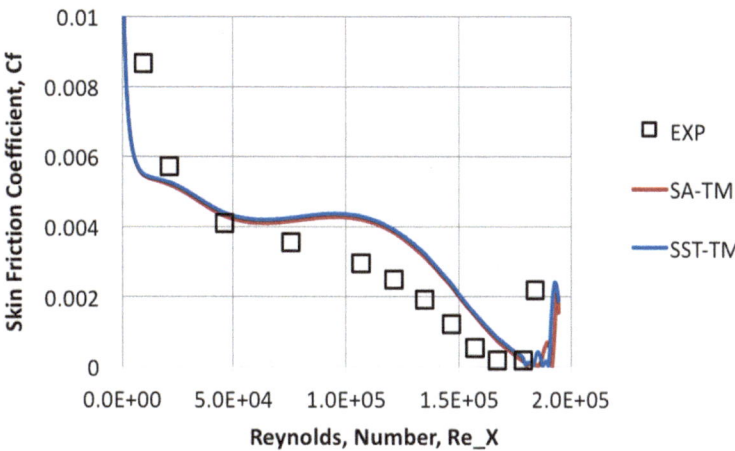

Fig. 4.10 Comparison of skin friction coefficient profiles for T3C4 case on flat plate

Another finding from the above computations is that the S-A based transition model has consistently predicted a slightly faster growth rate of turbulence, or a shorter transition length, when compared to the SST based transition model in both zero and non-zero pressure gradient cases. This is most likely due to the overestimate of the local free-stream turbulence intensity, which results in an overestimation of the F_{length} value in the S-A based transition model.

The last case in the T3C series is T3C4 ($Tu_\infty = 3.0\%$), which involves a separation-induced transition and reattachment of turbulent flow in the rear portion of the flat plate as shown in Fig. 4.10. The experimental data showed that the boundary layer separates on the flat plate at $x = 1.295$ m and reattaches at $x = 1.395$ m. In the computations, both S-A and SST transition models use the same inlet free-stream turbulence intensity and ambient turbulent eddy viscosity listed in Table 4.1. However, a local free-stream turbulence intensity evaluated based on Eq. (3.30) is used in the S-A based transition model. Figure 4.10 shows the computed skin friction coefficients in comparison with the experimental data, where nearly identical results are obtained using the S-A and SST transition models. However, both methods have predicted slightly delayed transition onsets than that observed in the wind tunnel.

4.3 Two-Dimensional Airfoils

Four two-dimensional airfoils are presented in this section for further validations of the S-A and SST-based transition models: Aerospatiale-A airfoil (Haase and Chaput 1993, Gendre 1992), VA-2 supercritical airfoil (Mateer and Manson 1993, 1996), S809 wind turbine airfoil (Somers 1997), and NACA 4412 airfoil (Coles and Wadcock 1979). These airfoils have been the subject of a large number of validations for transition and turbulence models in the CFD community (Haase et al. 1997; Langtry 2006). In particular, three of the above airfoils (Aerospatiale-A, S809, and NACA4412) involve flow separations or separation-induced transitions on the airfoils, which is ideal for examining the capability of these transition models to predict complicated transition phenomena in practical engineering applications. Inflow conditions for all two-dimensional airfoils are given in Table 4.2.

Computational meshes are generated using the unstructured mixed element grid topology. About 400 points are placed on the airfoil surface along chord-wise direction, and the mesh is refined at the leading and trailing edges of the airfoils. In the normal direction to the surface, the y^+ value for the nearest grid point off the wall is about one based on the corresponding reference Reynolds number for each case. A growth ratio of 1.1 is used to ensure that enough boundary layer nodes are

Table 4.2 Inflow conditions for two-dimensional airfoils

Test case	Inlet mach number	Inlet turbulence intensity (%)	Chord Re number (10^6)	Density (kg/m^3)	Molecular viscosity (kg/ms)
Aerospatiale A	0.15	0.05	2.1	1.2	1.8×10^{-5}
VA-2 super	0.2	0.5	0.6–6.0	1.2	1.8×10^{-5}
S809 wind turbine	0.1	0.05	2.0	1.2	1.8×10^{-5}
NACA 4412	0.009	0.0086	1.52	1.2	1.8×10^{-5}

Table 4.3 Comparison of computational costs for the Aerospatiale-A airfoil

Different models	Number of transport equations	CPU time (s)	RAM (GB)
SA turbulence model	1	5.22	2.17
SST turbulence model	2	6.31	2.25
SA-TM transition model	3	7.06	2.36
SST-TM transition model	4	8.27	2.45

generated to properly capture the development of the flow transition phenomena. The upstream and downstream boundaries are placed 20 chord lengths away from the airfoil to avoid the boundary reflection.

Computational cost is another important element in evaluating the practicality of turbulence models. Table 4.3 shows the typical computational costs in terms of CPU time and memory requirements on a per time step basis for the above two-dimensional airfoils. Comparing to the standard one-equation S-A turbulence model, the two-equation SST model costs about 20 % more in CPU time. The three and four-equation S-A and SST transition models add 35 % more cost to the respective S-A and SST turbulence models.

4.3.1   Aerospatiale-A Airfoil

The Aerospatiale-A airfoil (Haase et al. 1997), shown in Fig. 4.11, was designed at Aerospatiale in the ONERA F1 wind tunnel at an angle of attack of 13.1°. This experiment has been of great interest because there were no trips in the boundary layer measurement, and thus has been extensively used for the transition model validations. The chord length of this airfoil is 0.6 m, the free-stream Mach number is 0.15, and the Reynolds number is 2.1×10^6 based on the free-stream velocity.

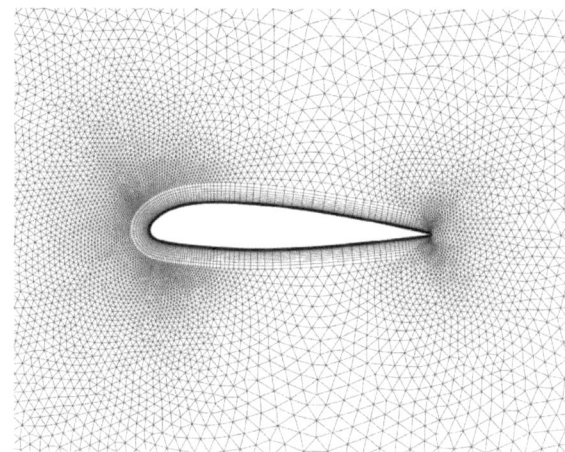

Fig. 4.11 Aerospatiale-A airfoil and computational mesh

The free-stream turbulence intensity is specified at 0.05 %. Both S-A and SST based transition models are evaluated for predicting the natural transition at the given inlet flow condition.

The skin friction coefficients calculated on the upper surface of the Aerospatiale-A airfoil are plotted in Fig. 4.12 for comparisons between the two models. In the experiment, the laminar boundary layer develops into turbulent flow and the transition onset occurs at 12 % of the chord length near the suction peak region, due to the laminar separation on the suction surface. It is seen that the standard SA and SST turbulence models, which treat the boundary layer as fully turbulent flow from the leading edge, are unable to capture the boundary layer transition phenomenon. However, both SA-TM and SST-TM transition models have captured the boundary layer transition successfully, where the predicted transition ranges from 12 to 15 % of the chord from the leading edge. The predicted skin friction coefficients exhibit a steep jump from zero value on the airfoil surface due to a separation-induced transition, which is consistent with the experiment.

The comparison of measured and predicted pressure coefficients for the Aerospatiale-A airfoil is shown in Fig. 4.13. The computed results using the standard turbulence models and the transition models both match the experimental data over the most of the airfoil surface. In the laminar flow region upstream of the boundary layer transition, the pressure coefficient profiles are slightly under-predicted using the standard SA and SST turbulence models. However, the SA-TM and SST-TM transition models are able to accurately capture the transition onset location in the boundary layer and thus enhance the prediction of the pressure coefficients in this region.

Comparisons of the lift and drag coefficients with the experimental data are given in Table 4.4. Although the SA-TM and SST-TM transition models slightly over-predict the lift coefficients, the drag coefficients are well predicted compared to the experimental data, and are within a 1.5 % difference of the measured value. In

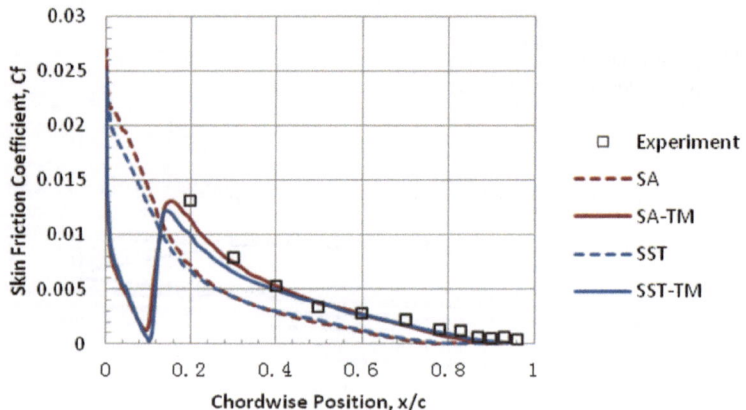

Fig. 4.12 Comparison of skin friction coefficients on the upper surface of the Aerospatiale-A airfoil

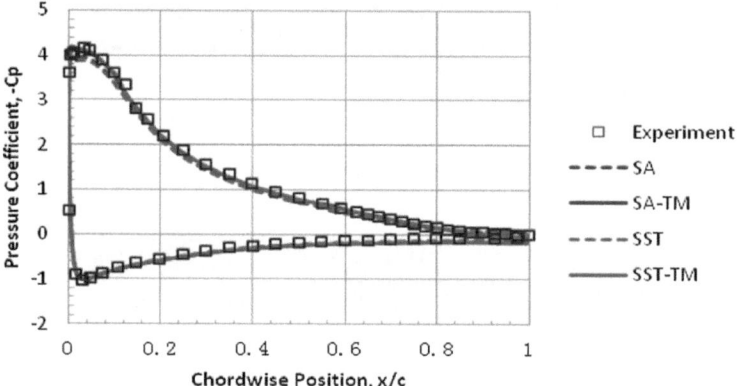

Fig. 4.13 Comparison of pressure coefficients on the Aerospatiale-A airfoil

Table 4.4 Comparison of lift and drag coefficients for the Aerospatiale-A airfoil

Values obtained from	Lift coefficient	Drag coefficient
Experiment	1.562	0.0208
SA turbulence model	1.54 (−1.4 %)	0.025 (+20.19 %)
SA-TM transition model	1.604 (+2.68 %)	0.0205 (−1.44 %)
SST turbulence model	1.546 (−1.02 %)	0.0253 (+21.63 %)
SST-TM transition model	1.608 (+2.94 %)	0.0206 (−0.96 %)

contrast, the standard turbulence models over-predict the drag coefficients by as much as 20 %.

The intermittency distributions obtained by SA-TM and SST-TM around the transition onset location ($x/c = 0.12$) are shown in Fig. 4.14. The intermittency (γ) equals zero in the boundary layer near the leading edge, which indicates that the laminar boundary layer has no eddy viscosity generated by the transport equation of the turbulence models. When the transition onset criterion is satisfied by the boundary layer flow, the intermittency starts to increase from zero to one, resulting in an increase of eddy viscosity correspondingly, as shown in Fig. 4.15.

4.3.2   VA-2 Supercritical Airfoil

The VA-2 airfoil is a supercritical airfoil that combines high lift and low drag features with a moderate loading. Experiments were performed at the NASA Ames High Reynolds Channel No. II by Mateer et al. (1993, 1996). A free-stream Mach number of 0.2 and free-stream turbulence intensity of 0.5 % are used in the CFD computations. This airfoil is used to test the S-A and SST transition models for

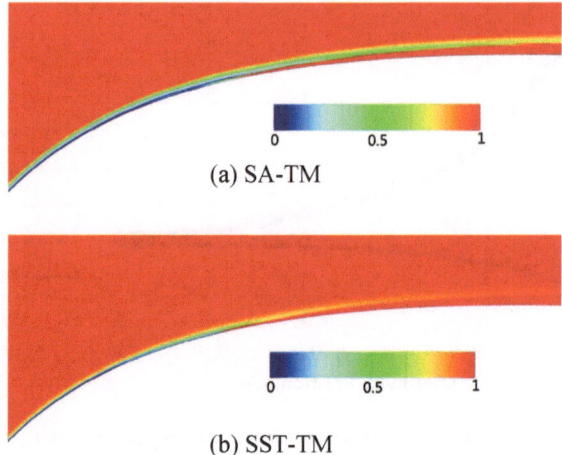

Fig. 4.14 Comparison of intermittency contours around the transition onset location of the Aerospatiale-A airfoil

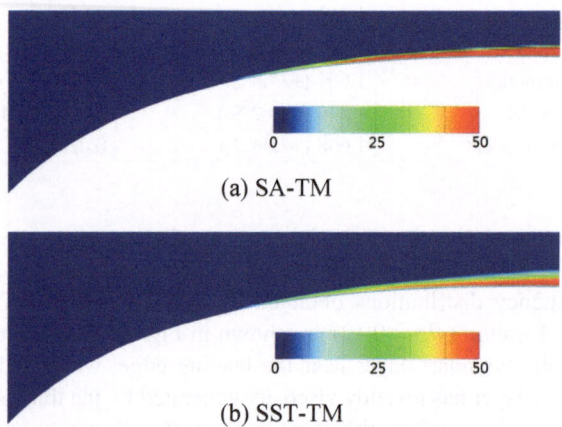

Fig. 4.15 Comparison of non-dimensional eddy viscosity contours around the transition onset location of the Aerospatiale-A airfoil

predicting transitions at different free-stream Reynolds numbers and angles of attack, in particular under the influence of strong pressure gradients. There are two flow conditions considered here, which are:

1. An angle of attack of $-0.5°$ at different Reynolds numbers (0.6×10^6, 2×10^6, and 6×10^6) based on the free-stream velocity and the chord length
2. A Reynolds number of 2×10^6 at different angles of attack ($-0.5°$, $3.5°$, $7.5°$ and $11.5°$)

Computational mesh for the VA-2 supercritical airfoil is shown in Fig. 4.16. The effect of the Reynolds number on the transition onset is shown in Figs. 4.17, 4.18 and 4.19, where the flow angle of attack is fixed at a negative 0.5°. As the Reynolds number increases, the transition onset locations on both upper and lower surfaces move from the downstream (close to the trailing edge) towards the upstream (close to the leading edge). Both SA-TM and SST-TM transition models have nicely captured the transition phenomena corresponding to the change of the Reynolds number. Because the VA-2 airfoil has a rather flat upper surface (similar to a flat plate without a pressure gradient), the transition onset location is primarily determined by the free-stream Reynolds number, which changes widely as shown in Figs. 4.17, 4.18 and 4.19. On the lower surface in the same figures, however, the adverse pressure gradient dominates the transition onset, and the transition location only moves slightly around the location of 45 % chord from the leading edge at different Reynolds numbers. These transition behaviors in response to the change of Reynolds number have been captured reasonably well by both SA-TM and SST-TM transition models.

Comparisons of the skin friction coefficient profiles at different angles of attack for a fixed Reynolds number of 2×10^6 are shown in Figs. 4.20, 4.21, 4.22 and 4.23. It is seen that as the angle of attack increases, the transition onset location moves towards the upstream on the upper surface. When the angle of attack is at 7.5° or above (Figs. 4.22 and 4.23), almost the entire upper boundary layer is turbulent flow. The skin friction coefficients predicted by the two transition models (SA-TM and SST-TM) are virtually the same as what are predicted by the standard turbulence models (SA and SST), which is to be expected because these transition models do not modify the physics in fully turbulent flows. On the lower surface, however, the transition onset location moves slightly towards downstream as the angle of attack increases, indicating a reduced level of turbulent activity in the boundary layer under strong flow acceleration environments (favorable pressure gradients).

Fig. 4.16 The VA-2 supercritical airfoil and computational mesh

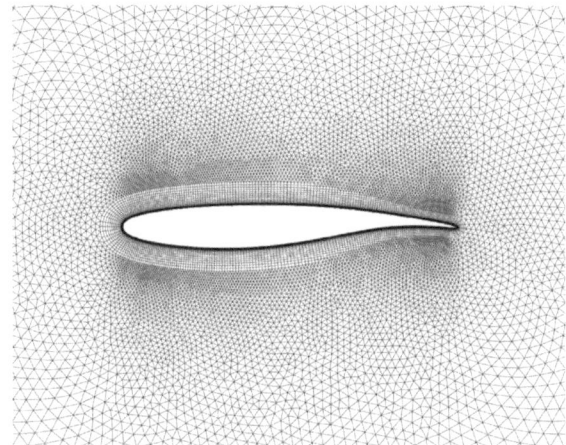

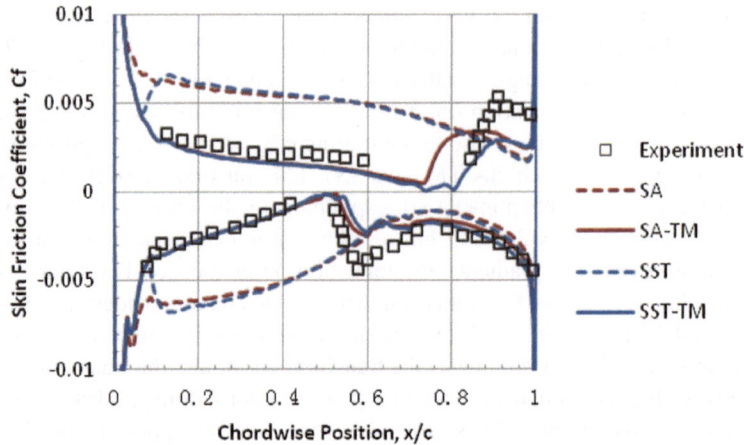

Fig. 4.17 Comparison of skin friction coefficients at $Re = 0.6 \times 10^6$ for the VA-2 supercritical airfoil

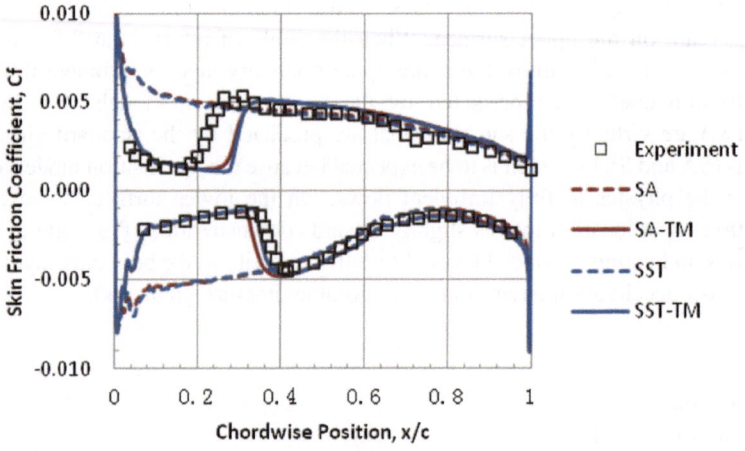

Fig. 4.18 Comparison of skin friction coefficients at $Re = 2.0 \times 10^6$ for the VA-2 supercritical airfoil

4.3.3 S809 Wind Turbine Airfoil

The S809 airfoil is a 21 % thick, 0.6 m chord length, laminar flow airfoil that was designed specifically for horizontal-axis wind turbines. An experiment was performed in the 1.8 by 1.25 m, low-turbulence wind tunnel at Delft University of Technology (Somers 1997), with the primary design objectives being to restrain maximum lift, be insensitive to roughness, and have a low profile drag in typical operating environments. Therefore, at a low angle of attack up to 5°, the laminar

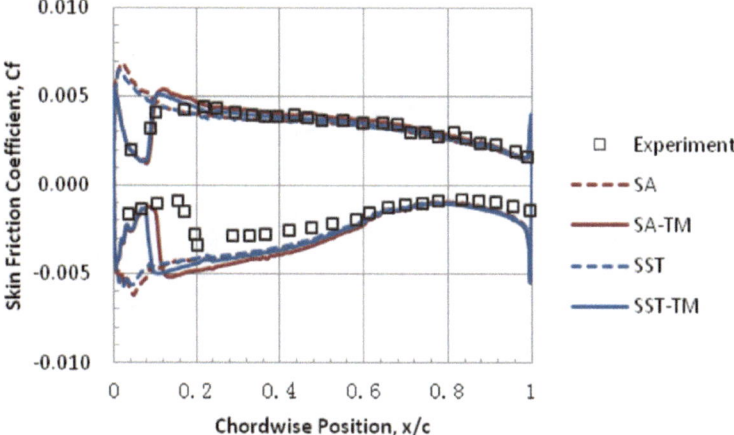

Fig. 4.19 Comparison of skin friction coefficients at $Re = 6.0 \times 10^6$ for the VA-2 supercritical airfoil

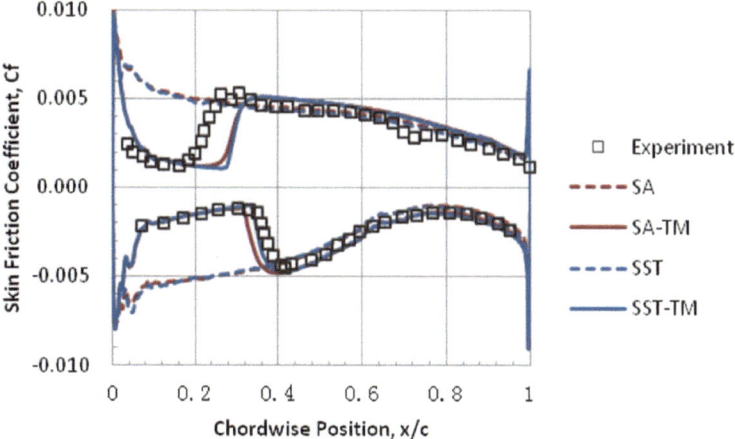

Fig. 4.20 Comparison of skin friction coefficients at $AoA = -0.5°$ for the VA-2 supercritical airfoil

boundary layer covers up to 50 % of the chord on both sides of the S809 airfoil. After that, the flow undergoes a laminar separation followed by a turbulent reattachment. Further increasing the angle of attack causes the transition point to move towards upstream and results in a small amount of turbulent flow separation at the trailing edge. The trailing edge separation expands to about 5–10 % of the chord at 9°, and nearly 50 % of the chord at 15°. As the angle of attack is increased to 20°, most of the upper surface is stalled.

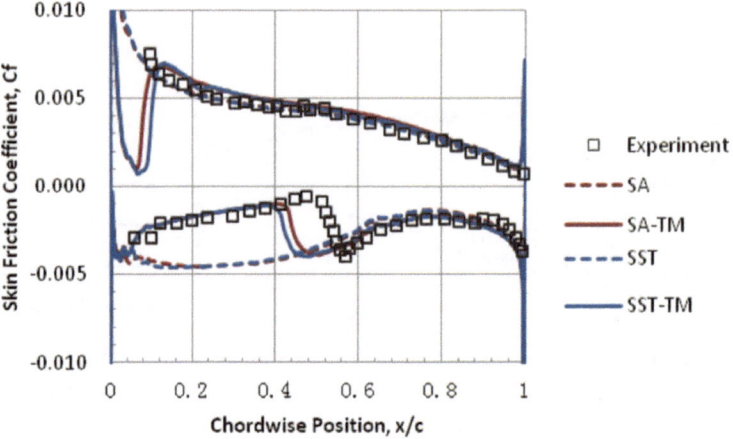

Fig. 4.21 Comparison of skin friction coefficients at *AoA* = 3.5° for the VA-2 supercritical airfoil

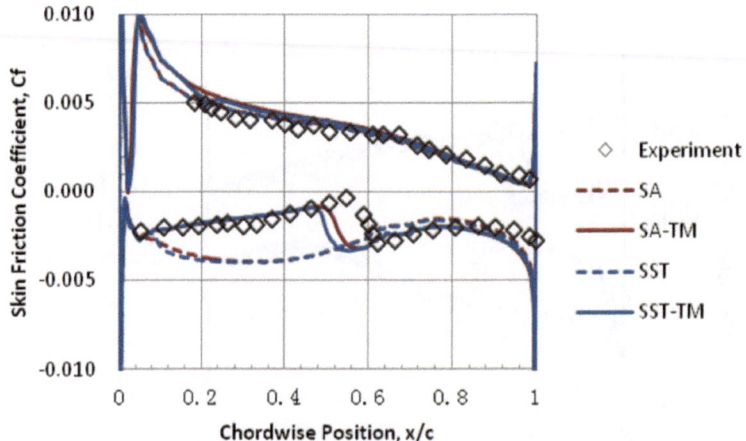

Fig. 4.22 Comparison of skin friction coefficients at *AoA* = 7.5° for the VA-2 supercritical airfoil

Computational mesh for the S809 wind turbine airfoil is generated and shown in Fig. 4.24. The Reynolds number is 2.0×10^6 based on the chord length. The free-stream turbulence intensity is 0.05 %, which would correspond to the turbulence level in a typical low-turbulence wind tunnel. The free-stream Mach number is not given in the experiment's report, but is specified as 0.1 in the CFD simulations. Detailed transition onset locations at different angles of attack can be found in the experimental report (Somers 1997).

Comparisons of predicted lift and drag coefficients along with the experimental data are shown in Figs. 4.25 and 4.26. The results show that predicted C_l and C_d coefficients (found from the SA-TM and SST-TM solutions) match well with the

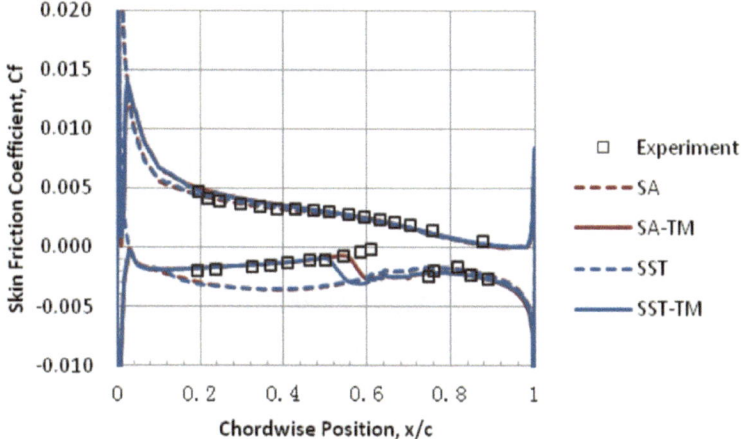

Fig. 4.23 Comparison of skin friction coefficients at $AoA = 11.5°$ for the VA-2 supercritical airfoil

Fig. 4.24 The S809 airfoil and computational mesh

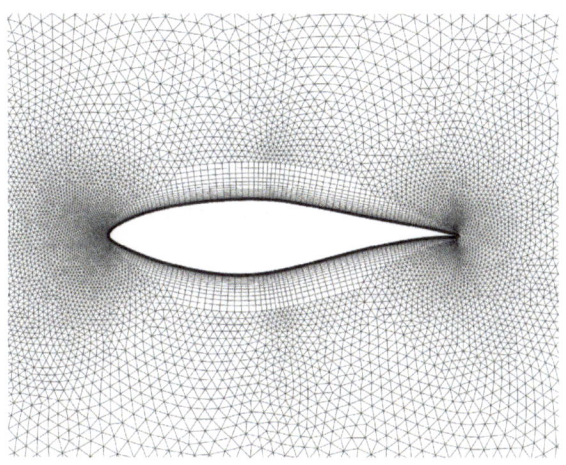

experimental data for the angle of attack below 9°. After this angle of attack, there are some discrepancies between the predictions and the wind tunnel measurements. As discussed above, an enlarged separation on the upper surface at the high angle of attack is the main source of errors in the CFD predictions, because no turbulence models yet are fully calibrated for predicting massively separated flows. Another source of errors is the three-dimensional effect in the wind tunnel test, which cannot be correctly captured in two-dimensional airfoil simulations. However, there are two interesting and worthwhile observations from this study. Firstly, after the flow separates at high angles of attack, both SA-TM and SST-TM transition models do not show distinct advantages over the standard SA and SST models in the lift and drag predictions. Secondly, it appears that the two-equation turbulence model, with

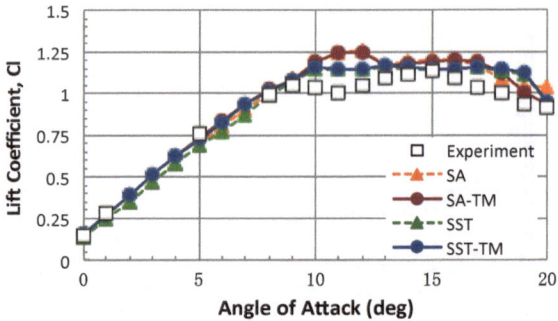

Fig. 4.25 Comparison of lift coefficients for the S809 airfoil

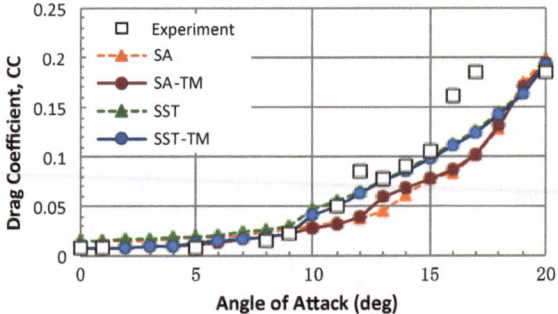

Fig. 4.26 Comparison of drag coefficients for the S809 airfoil

or without the transition modeling (SST or SST-TM), provides better predictions on the lift and drag coefficients than the one-equation turbulence model (SA or SA-TM), at least for the current separated flows as shown in Figs. 4.25 and 4.26.

The comparison of the predicted and measured transition onset locations for the S809 airfoil is shown in Fig. 4.27. The predictions from using SA-TM and SST-TM transition models generally agree well with the experimental data. In the case of a lower angle of attack (less than 5°), the location of the transition onset is around 50 % chord on both suction and pressure surfaces. As the angle of attack increases, the transition onset location on the suction surface moves towards upstream. However, predicted transition onset locations are slightly off at 5° or 6°. On the pressure surface, the adverse pressure gradient dominates the boundary layer transition process, and the transition onset location is at around 50 % chord for all test conditions. In general, predicted transition onset locations are similar using both SA-TM and SST-TM transition models.

The pressure coefficient profiles predicted at the angle of attack of 1°, 9°, 14° and 20° on the S809 airfoil surface are compared with the measurements in Figs. 4.28, 4.29, 4.30 and 4.31. In general, all predicted results using different turbulence or transition models agree well with the experimental data at low or

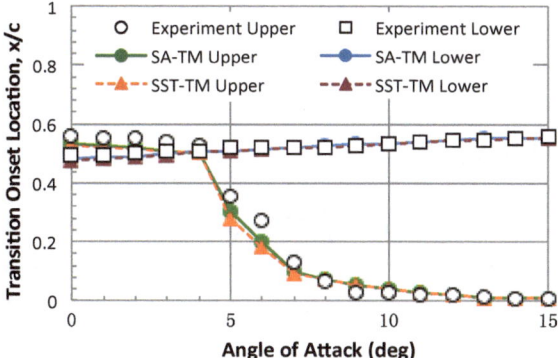

Fig. 4.27 Comparison of transition onset locations for S809 airfoil

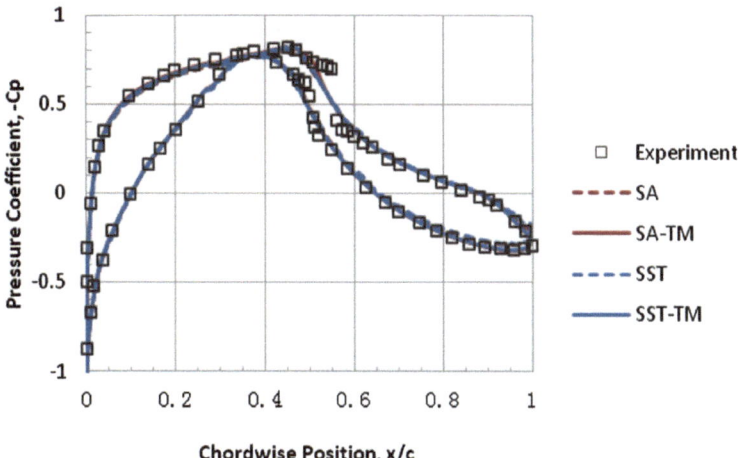

Fig. 4.28 Comparison of pressure coefficients at $AoA = 1°$ for the S809 airfoil

moderate angles of attack, as shown in Figs. 4.28, 4.29, 4.30 and 4.30. At a high angle of attack of 20°, there are some discrepancies at the leading edge suction surface where the flow stalls, which is shown in Fig. 4.31. Unlike the skin friction, the pressure profile on the airfoil surface is less sensitive to the transition modelling methods, but can be severely affected by flow separations or stall.

4.3.4 NACA 4412 Airfoil

The final two-dimensional validation case presented in this chapter is the NACA 4412 airfoil (Coles and Wadcock 1979), which is a well-known airfoil of the

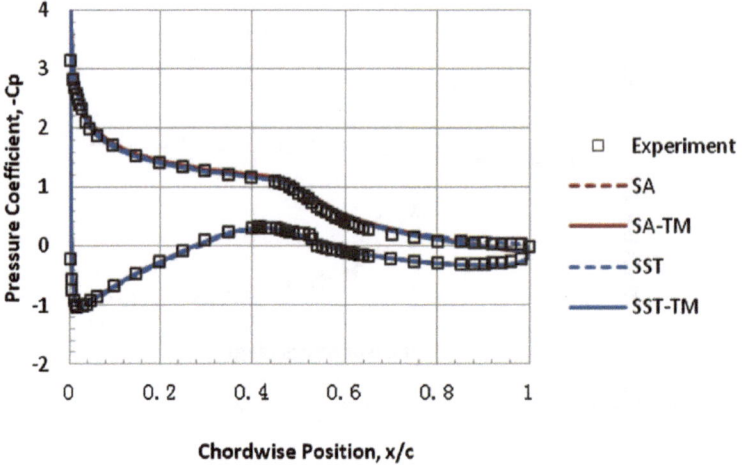

Fig. 4.29 Comparison of pressure coefficients at *AoA* = 9° for the S809 airfoil

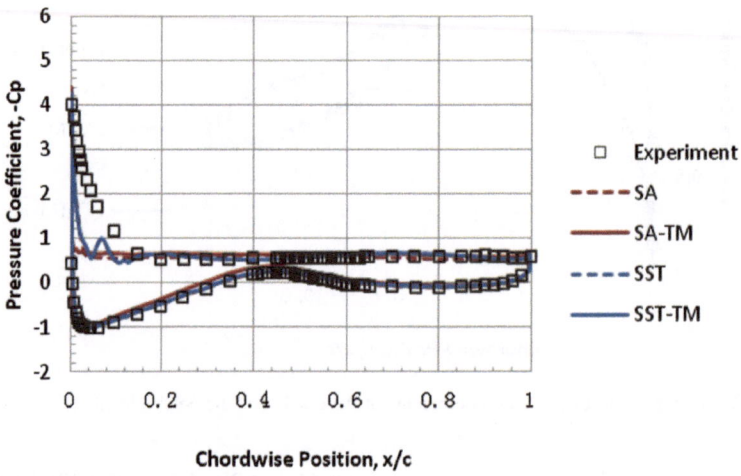

Fig. 4.30 Comparison of pressure coefficients at *AoA* = 14° for the S809 airfoil

NACA four-digit series. The first two digits indicate the maximum camber and location (44 means 4 % chord length maximum camber at 4 % chord line from the leading edge), and the last two digits indicate the maximum thickness in percentage of the mean aerodynamic chord (12 % here). Both upper and lower surface boundary layers were tripped in the experiment (2.5 % chord upper surface and 10.3 % chord lower surface) performed by Coles and Wadcock (1979). The experimental data include both the mean velocity and the Reynolds shear stress profiles in the separation zone at the airfoil's trailing edge. Since the flow is tripped,

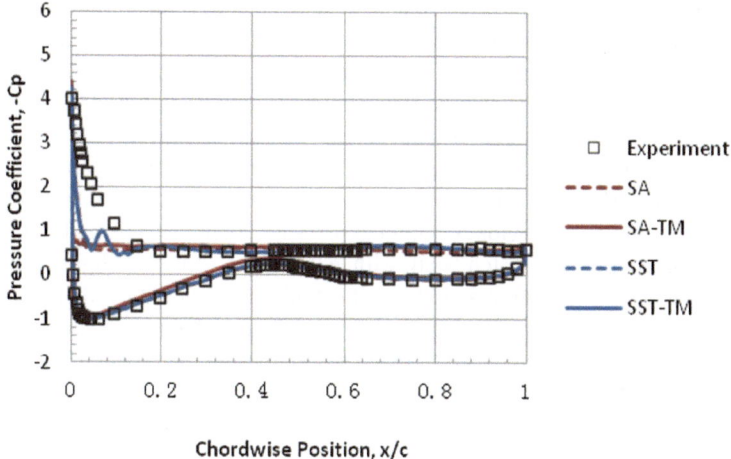

Fig. 4.31 Comparison of pressure coefficients at *AoA* = 20° for the S809 airfoil

this case is not used for validating the transition models, but instead for validating the separation correction method (Eq. 3.32) introduced in Chap. 3. The Reynolds number is 1.52×10^6 in the computation (based on the airfoil chord length) and the Mach number is 0.09 with an angle of attack of 13.87°. Figure 4.32 shows the NACA 4412 airfoil profile and the computational mesh. A refined grid resolution in the trailing edge is generated in order to predict the mean velocity distributions and investigate the turbulent shear stress profiles within the separation zone. The mesh size for the NACA 4412 airfoil is about 1.96 million nodes.

Fig. 4.32 The NACA 4412 airfoil and computational mesh

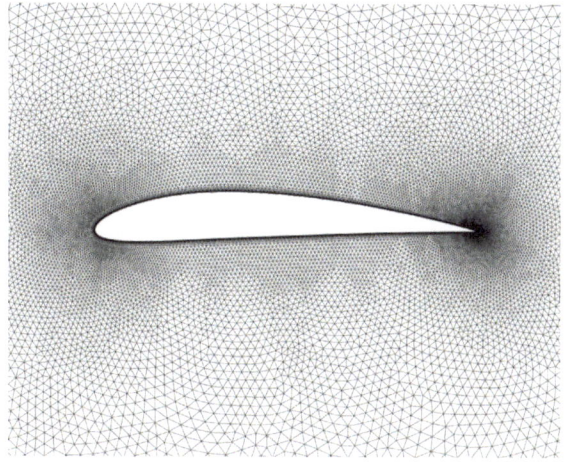

Computations are performed using both S-A and SST full turbulence models, where the production term of the eddy viscosity in both models is modified by the separation correction (Eq. 3.32). Shown in Fig. 4.33 is the comparison of the measured and computed pressure coefficient distributions on the NACA 4412 air-foil surface at 13.87° angle of attack. On the upper surface, the pressure is decreased sharply from the leading edge. The trailing edge separation exists in the rear part of the airfoil upper surface from $x/c = 0.8$ to 1, where the magnitude of C_p is almost constant in the region. Computed pressure coefficients using both S-A and SST turbulence models match well with each other, but are slightly off when compared with the experimental data. It should be noted that there are some uncertainties in the experimental data as reported by Coles and Wadcock (1979).

Computed velocity profiles are assessed by comparing them with the wind tunnel data in the trailing edge separation zone. There are six measurement locations ($x/c = 0.6753, 0.7308, 0.7863, 0.8418, 0.8973$ and 0.9528) in the normal direction to the airfoil near the trailing edge (Fig. 3.13), where detailed boundary layer profiles were measured in the wind tunnel. As shown in Figs. 4.34 and 4.35, computed stream-wise mean velocity profiles (U/U_{ref}) show good conformity with the wind tunnel measurements using both S-A and SST turbulence models. The computed mean velocity profiles in the normal direction to the airfoil surface (V/U_{ref}) show acceptable agreement with the measurements, as shown in Figs. 4.36 and 4.37. The normalized shear stresses $U'V'/U_{ref}^2$, on the other hand, are found to be noticeably under-predicted when compared with the wind tunnel measurements even with the separation correction method, which are shown in Figs. 4.38 and 4.39. However,

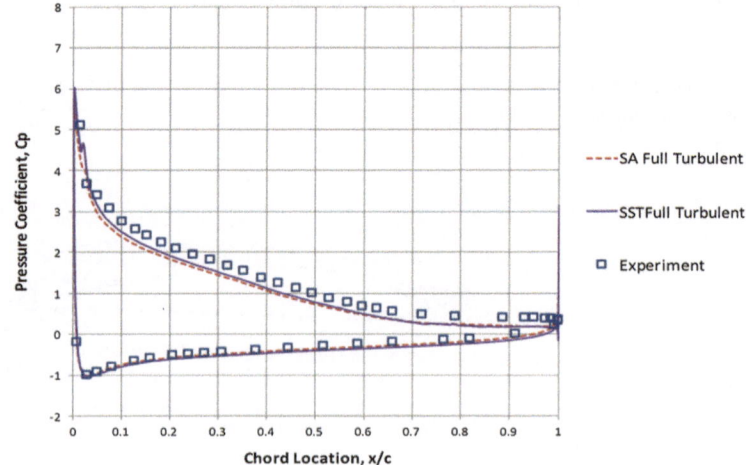

Fig. 4.33 Comparison of pressure coefficients at $AoA = 13.87°$ for the NACA 4412 airfoil

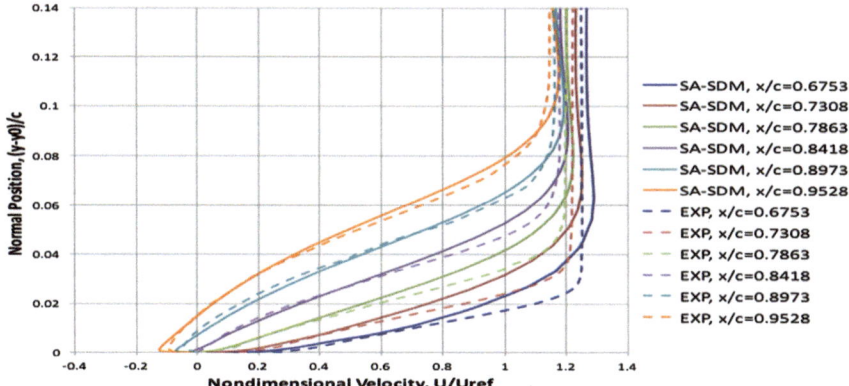

Fig. 4.34 Comparison of non-dimensional velocity in x direction obtained by SA-SDM model

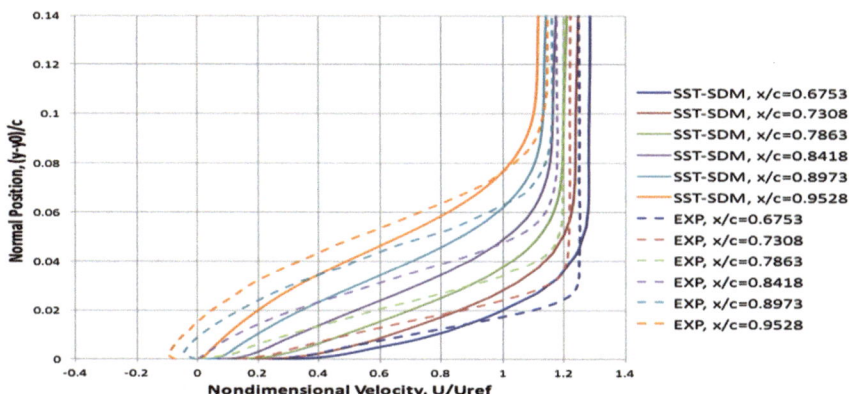

Fig. 4.35 Comparison of non-dimensional velocity in x direction obtained by SST-SDM model

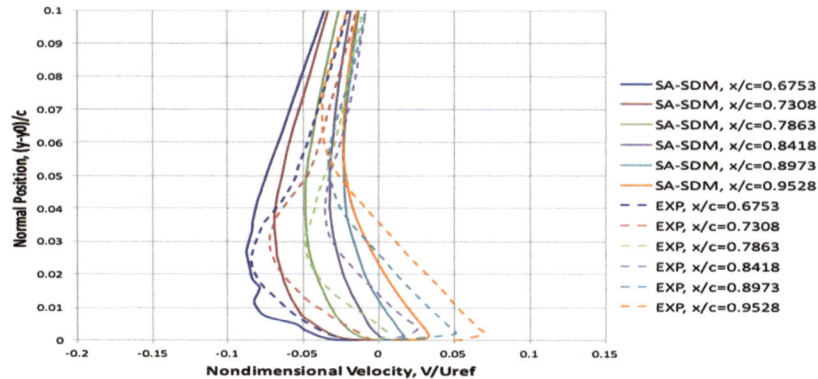

Fig. 4.36 Comparison of non-dimensional velocity in y direction obtained by SA-SDM model

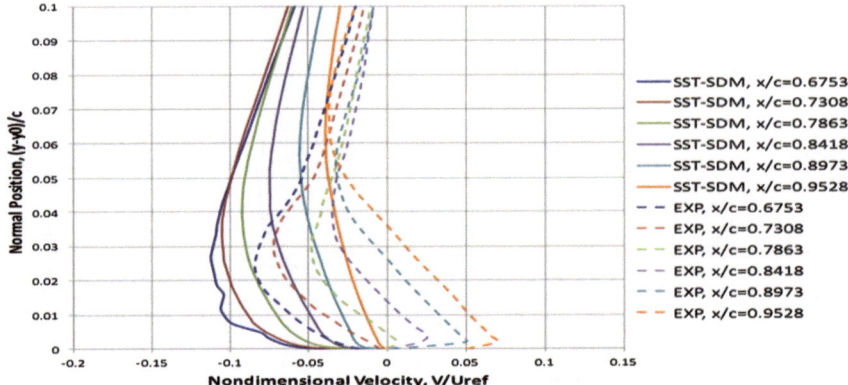

Fig. 4.37 Comparison of non-dimensional velocity in *y* direction obtained by SST-SDM model

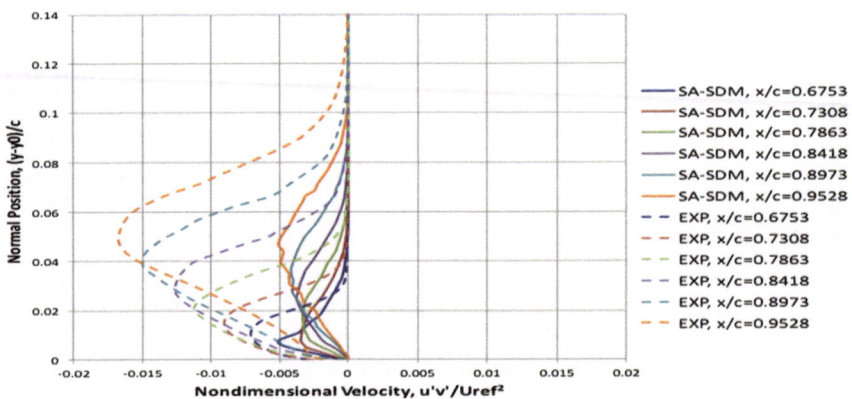

Fig. 4.38 Comparison of non-dimensional shear stress obtained by SA-SDM model

the computed mean velocity profiles would not even match with the experimental data if the separation correction was not used. This is consistent with the computational results obtained by other CFD codes that are published at NASA Langley Research Center Turbulence Modeling Resource website (http://turbmodels.larc. nasa.gov/naca4412sep_val.html). These results further demonstrate the deficiency of the RANS modeling approaches in predicting flows near separations (Aupoix et al. 2011).

Numerical experiments using the separation correction method indicate that, although increasing the separation correction factor (S_2) further would boost the level of turbulent shear stresses in the separation zone, the mean velocity profiles are also altered if the correction to the separated flows is too large. It remains challenging to achieve an optimal solution with both corrected turbulent shear stresses and mean velocity profiles in the separated zone. The behavior of the

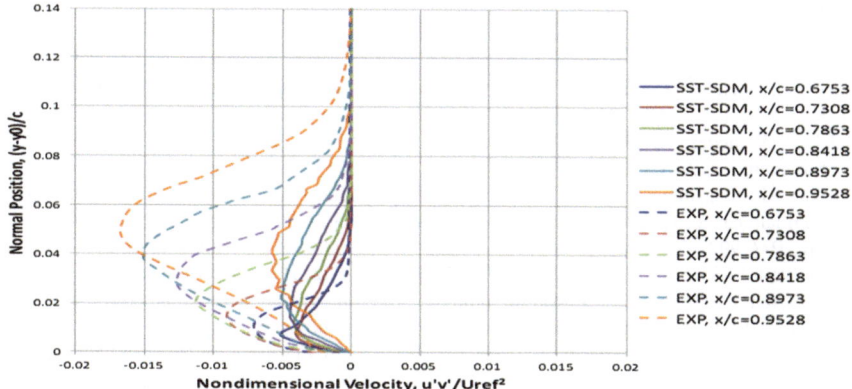

Fig. 4.39 Comparison of non-dimensional shear stress obtained by SST-SDM model

current separation correction method (Eq. 3.32) suggests that an enhancement or alternative method may be pursued. Although challenging, this may be necessary in order to address this fundamental issue for the RANS modeling methods near flow separations.

4.4 Summary

In this chapter, several two-dimensional benchmark flow validation cases are discussed for the S-A and SST-based transition models by comparing the numerical results with the wind tunnel experimental data. The transition models are assessed for predicting the boundary layer transition phenomena over the flat plates at different free-stream turbulence intensities, with and without the influence of the pressure gradient, based on the European Research Community on Flow, Turbulence and Combustion (ERCOFTAC) T3 series of experimental tests. The sensitivity studies of the SA-TM and SST-TM transition models on the free-stream Reynolds number and the angle of attack are also investigated based on several two-dimensional airfoils. The computational results show that the current transition models correctly respond to various flow conditions and provide acceptable predictions of the transition onset locations for the problems evaluated. In addition, a new separation correction, called the Stall Delay Method (SDM), is also investigated for predicting flows involving separations.

In the next chapter, both S-A and SST based transition models, including SDM, will be further investigated for predicting the aerodynamic performance and flow field of three-dimensional realistic helicopter or tilt rotors in hover, to support the rotor aerodynamic analysis and design optimization.

References

Aupoix B, Arnal D, Bezard H et al (2011) Transition and turbulence modeling. J Aerosp Lab, issue 2, Mar 2011

Coles D, Wadcock AJ (1979) Flying-hot-wire study of flow past an NACA 4412 airfoil at maximum lift. AIAA J 17(4):321–329

Dhawan SJ, Narasimha R (1958) Some properties of boundary layer flow during the transition from laminar to turbulent motion. J Fluid Mech 3(4):418–436

Gendre P (1992) Maximum lift for a single-element airfoils-experimental results A-airfoil ("CER-FACS" contribution in task 2.1 to EUROVAL). In: Haase W, Brandsma F, Elsholz E et al (eds) EURPVAL-A European initiative 184 on validation of CFD codes, notes on numerical fluid mechanics, vol 42. Vieweg Verlag, Braunschweig

Haase W, Brandsma F, Elsholz E et al (1993) EURPVAL-A European initiative on validation of CFD codes. In: Haase W, Chaput E, Elsholz E (eds) Notes on numerical fluid mechanics and multidisciplinary design. Vieweg Verlag, Braunschweig

Haase W, Chaput E, Elsholz E et al (1997) ECARP—European computational aerodynamics research project: validation of CFD codes and assessment of turbulence models. In: Haase W, Chaput E, Elsholz E (eds) Notes on numerical fluid mechanics, Viehweg, Wiesbaden

Langtry RB (2006) A correlation-based transition model using local variables for unstructured parallelized CFD codes. Ph.D. Dissertation, University of Stuttgart. http://elib.uni-stuttgart.de/opus/volltexte/2006/2801/

Langtry RB, Menter FR (2009) Correlation-based transition modeling for unstructured parallelized computational fluid dynamics codes. AIAA J 47(12):2894–2906

Mateer GG, Seegmiller HL, Hand LA et al (1993) An experimental investigation of a supercritical airfoil at transonic speeds. In: NASA-TM-103933, 1993

Mateer GG, Monson DJ, Menter FR (1996) Skin-friction measurements and calculations on a lifting airfoil. AIAA J 34(2):231–236

Menter FR (1994) Two-equation eddy-viscosity turbulence models for engineering application. AIAA J 32(8):1598–1605

Savill AM (1993a) Some recent progress in the turbulence modeling of by-pass transition. In: So RMC, Speziale CG, Launder BE (eds) Near-wall turbulent flows. Elsevier, Amsterdam, pp 829–848

Savill AM (1993b) Evaluating turbulence model predictions of transition. In: Advances in turbulence IV. Springer Netherlands, pp 555–562

Schubauer GB, Klebanoff PS (1956) Contribution on the mechanics of boundary layer transition. In: NACA TN 3489, Jan 1956

Somers DM (1997) Design and experimental results for the S809 airfoil. In: National Renewable Energy Lab, SR-440-6918, Jan 1997

Spalart PR, Allmaras SR (1994) A one-equation turbulence model for aerodynamic flows. La Rech Aerospatiale 1:5–21

Suluksna K, Dechaumphai P, Juntasaro E (2009) Correlation for modeling transitional boundary layers under influences of free stream turbulence and pressure gradient. Int J Heat Fluid Flow 30(1):66–75

Suzen YB, Xiong G, Huang PG (2002) Predictions of transitional flows in low-pressure turbines using intermittency transport equation. AIAA J 40(2):254–266

Chapter 5
Applications for 3-D Rotors

Abstract The S-A and SST-based transition models are used to predict the aerodynamic performance of three realistic helicopter and tiltrotor blades: the XV-15 proprotor, the Joint Vertical take-off/landing eXperimental rotor, and the S-76 conventional rotor. The role of transition modeling in rotor hover predictions is investigated based on comparative studies of different models, and predicted transition phenomena are validated with the available experimental data. The focus of this chapter is to address three major issues pertinent to rotor hover predictions: (1) effect of transition modeling on the rotor aerodynamic prediction, (2) rotor predictions at high thrusts or collective angles, and (3) rotor flow characteristics and its influence on aerodynamic performance.

5.1 XV-15 Proprotor[1]

The Bell XV-15 tiltrotor is a research aircraft used to demonstrate the concept of high-speed performance relative to conventional helicopters. A complete history of the XV-15 tiltrotor research aircraft from concept to flight was reported by Maisel et al. (2000). The XV-15 program was launched in 1971 at NASA Ames Research Center. After competition of the prototype designs, the Bell Model 301 was selected for further development. In 1983, Bell Helicopter and Boeing Vertol teamed up to submit a bid for an enlarged version of the XV-15 for the DoD's Joint-service Vertical take-off/landing eXperimental (JVX) aircraft program. The Bell-Boeing team won a preliminary design contract that year, which later led to the successful development of the Bell-Boeing V-22 Osprey (Maisel et al. 2000).

Throughout the past decades, many researchers have investigated the aerodynamic performance of the XV-15 proprotor. Figure 5.1 shows the full-scale XV-15 tiltrotor research aircraft in the NASA Ames 40 × 80 ft wind tunnel. These highly twisted rotor blades are designed for high speed performance in axial flight, which

---

[1]Original work was published in the Proceedings of the American Helicopter Society (AHS) International 72nd Annual Forum and Technology Display, 16–19 May 2016, West Palm Beach, Florida, U.S.A.

© The Author(s) 2017   83
C. Sheng, *Advances in Transitional Flow Modeling*, SpringerBriefs
in Applied Sciences and Technology, DOI 10.1007/978-3-319-32576-7_5

Fig. 5.1 Bell XV-15 tiltrotor
research aircraft in NASA
Ames 40 × 80 ft wind tunnel
(*Source* http://www.nasa.gov/
centers/ames/images/)

represents the domain of tiltrotor in airplane and helicopter mode. Felker et al.
(1985) conducted detailed performance and load measurements for the full-scale
XV-15 rotor. Wadcock and Yamauchi (1998), as well as Wadcock et al. (1999)
measured the skin frictions of the full-scale XV-15 rotor at two tip Mach numbers
using the oil-film interferometric skin friction technique. Their wind tunnel measurements showed the leading-edge laminar separation bubbles at high rotor collective pitch angles (Wadcock et al. 1999), as illustrated in Fig. 5.2. Leading edge
separation bubbles are common on many rotorcraft blades, which cause the
so-called separation-induced transition (see Chap. 1). In addition, due to highly
twisted nature of the XV-15 rotor blades, there were extensive reversed flows at the
inboard sections ranging from the middle chord to the trailing edge of the blade,
which was observed in the wind tunnel tests, especially at high rotor collective
angles (greater than 10°). The wind tunnel test image in Fig. 5.3 shows the reversed
flows of the XV-15 rotor flow field at a high thrust level ($\theta = 16.6°$, $C_T = 0.0145$).

Several researchers also conducted computational investigations for the XV-15
tiltrotor. Kaul and Ahmad (2011), as well as Kaul (2012) studied the effect of the
inflow boundary conditions and various turbulence models on the hovering XV-15
rotor flow field using a CFD code called OVERFLOW2 (Nichols et al. 2006). Yoon
et al. (2014) also performed similar simulations of the XV-15 rotor in hover. In both
simulations, the Spalart's Detached Eddy Simulation (DES) method (Spalart 2009)
was used in conjunction with the Spalart-Allmaras (S-A) and Menter's Shear Stress
Transport (SST) turbulence models. Their computational results showed varying
agreements between the predicted and measured figure of merit (FM). Predictions
also showed disagreements with the experimental skin friction profiles (Wadcock
et al. 1999) in the region near the blade's leading edge, because neither of their
computations have the capability to predict the transitional flow phenomenon
observed in the experiment. The XV-15 rotor prediction using a $\kappa - \omega - \gamma$
three-equation transition model was recently performed by Zhao et al. (2014). In
addition, Sheng et al. (2016b) performed a comparative study for the XV-15 rotor
using the U^2NCLE (Sheng 2011) and Helios (Wissink et al. 2012) CFD codes,

Fig. 5.2 XV-15 wind tunnel fringe patterns at high thrust, θ = 14.1° (*photo* courtesy of Wadcock and Yamauchi 1998)

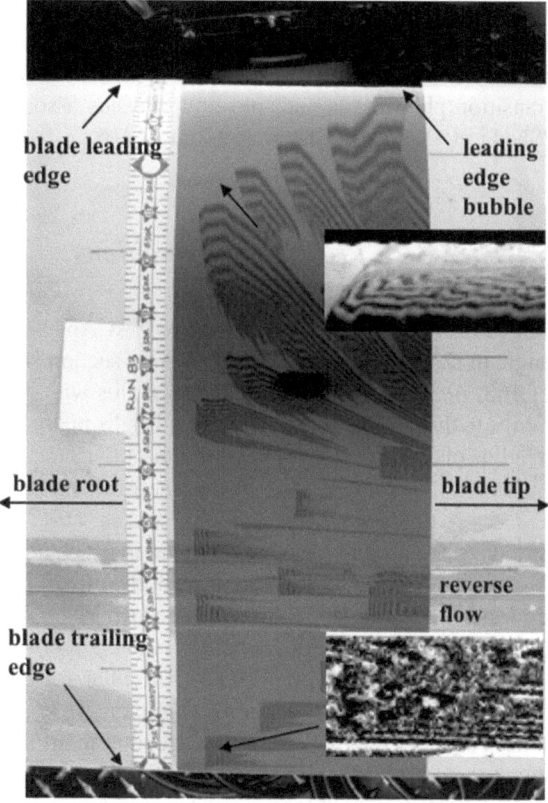

Fig. 5.3 XV-15 image of turf motion over the inboard blade at θ = 16.6° (*photo* courtesy of Wadcock and Yamauchi 1998)

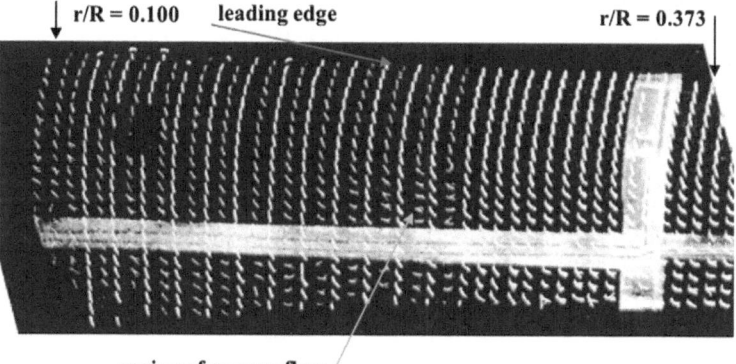

where the SA-TM transition model was used in U²NCLE and the standard Spalart-Allamars turbulence model was used in Helios. Their computational results indicated that a transition modeling capability is required in order to capture the transition phenomena and true flow physics associated with the XV-15 proprotor (Sheng et al. 2016b; Sheng and Zhao 2016).

5.1.1 XV-15 Profile and Conditions

The XV-15 rotor blade comprises five NACA64 series profiles with a 2.5° pre-cone angle in hover. The inboard aerodynamic section starts at 9.1 % radius with a chord of 16.6 in., linearly tapering to 25 % radius with a chord of 14 in., and from 25 % radius to the tip, a constant chord of 14 in. is maintained. This three-bladed rotor has a radius of 150 in., with a blade structural twist angle of −37.35° from the root to the tip. A sketch for the XV-15 blade planform and the blade twist profile is shown in Fig. 5.4, based on the report of Felker et al. (1985). The full-scale rotor tests were measured at different rotor RPMs, with tip Mach numbers ranging from 0.56 to 0.73. The computational investigations are performed at two tip Mach numbers: the design tip Mach number of 0.69 and a reduced tip Mach number of 0.56. The focus of this investigation is to conduct detailed validations of the S-A and SST based transition models described in this book, in particular for capturing the separation-induced transition as well as the inboard flow separation observed on the XV-15 wind tunnel tests. Table 5.1 provides the XV-15 geometric information, rotor tip chord Reynolds number, and tip Mach numbers used for the hover computations.

Since the flow field for an isolated XV-15 rotor in hover is axisymmetric around the rotor axis, computations are performed in computational domain that consists of a single blade only, as shown in Fig. 5.5. An axisymmetric boundary condition is

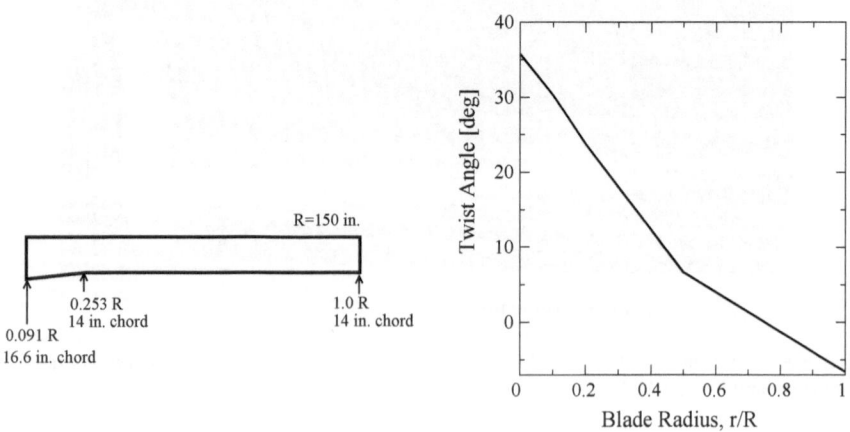

Fig. 5.4 XV-15 rotor blade platform and twist distribution

Table 5.1 A summary of XV-15 full-scale rotor geometry and flow conditions

Rotor radius (in.)	150	150
Solidity	0.0891	0.0891
Tip chord (in.)	14	14
Taper (tip/root chord)	0.7856	0.7856
Tip Mach number	0.56	0.69
Tip speed (ft/s)	628.32	771.54
Chord reynolds number	4.6×10^6	5.6×10^6
Collective pitch (deg.)	10, 12, 14, 16.6	0, 3, 6, 8, 10, 12, 14, 16, 16.6

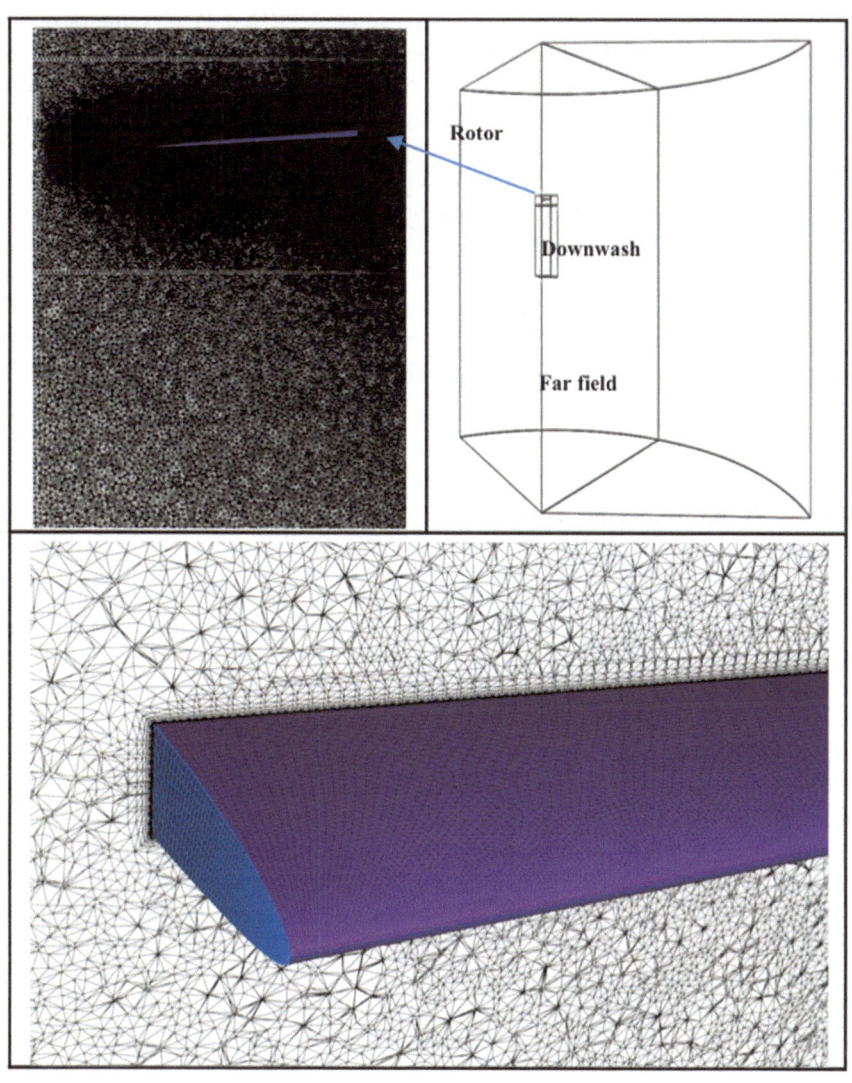

Fig. 5.5 Single blade computational domain in 120° sector (*top*) and boundary layer grids at the blade tip (*bottom*)

applied to the periodic boundaries in the computational domain. This provides significant savings in computational time and resources. For the present three-bladed XV-15 rotor, the mesh size is 23.7 million points for the single blade, equivalent to a mesh size of 71.1 million points for the entire XV-15 rotor, if all three blades are included in the computational domain.

Previous hover computations indicated that sufficient surface and volume mesh resolutions are essential for accurate predictions of rotor performance and capturing the transition onset location, especially at high blade collective angles (Sheng et al. 2016b). A refined surface mesh resolution at the leading and trailing edges of the blade is required to correctly capture the near-body inflow wake and the transition onset. As a rule of thumb, the maximum surface point spacing on the blade should be about 1–2 % of the tip chord length (Sheng et al. 2016a). A y^+ value of one is usually set in the normal direction of the blade surface in order to resolve the viscous boundary layer and capture the transition phenomenon. Figure 5.5 also shows the boundary layer mesh and the surface resolution in the blade tip region.

A method of mesh deformation (Sheng and Allen 2013) is applied here for performing the rotor collective angle sweeps. As the XV-15 hover performance is predicted over the entire collective pitch angles, this technique avoids regenerating the computational mesh by deforming or moving the existing rotor surface and volume meshes into a new position during the collective angle sweeps. The mesh deformation, combined with the periodic boundary condition for a single blade computational domain, provides a cost effective way to obtain an engineering solution for the isolated rotor aerodynamic performance in hover or axial flight conditions.

5.1.2   Hover Performance

The XV-15 rotor computations performed by Sheng et al. (2016b) indicated that similar hover performance results were obtained using both the U^{2}NCLE and Helios codes, where the SA-TM transition model and standard S-A full turbulence model were used in these computations, respectively. The effects of the transition modeling on the XV-15 hover performance are further discussed in this chapter using the SA-TM and SST-TM transition models. Figures 5.6 and 5.7 show comparisons of the FM and C_P/σ versus C_T/σ coefficients between the computations and measurements at the tip Mach number of 0.69 (Felker et al. 1985). In general, predicted results by both SA-TM and SST-TM transition models match very well with the experimental data, except for the high blade collective angle greater than 14° (or $C_T/\sigma > 0.17$). The average difference between the predicted and measured FM is about 1.5 %, indicating an excellent correlation between the computed and measured hover performance over the entire rotor collective range. It should be noted that the Stall Delay Method (SDM) introduced in Chap. 3 is applied here in order to prevent premature flow separations at high blade collective angles.

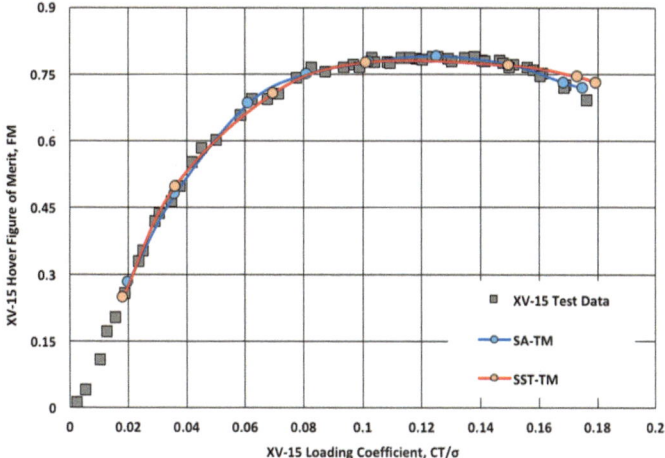

Fig. 5.6 Comparison of measured and predicted FM versus C_T/σ using SA and SST based transition models, $M_{tip} = 0.69$ (Felker et al. 1985)

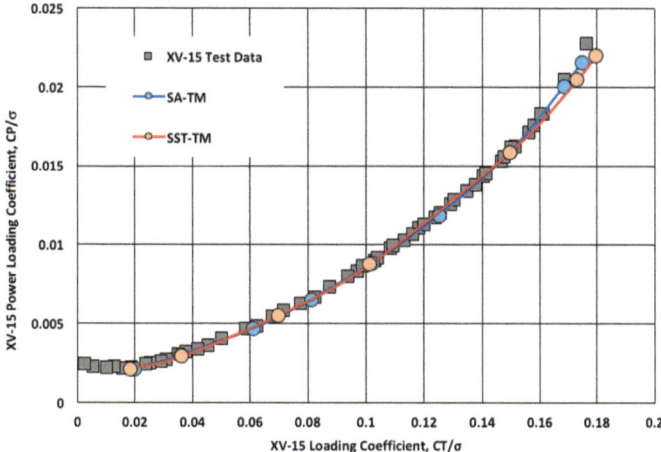

Fig. 5.7 Comparison of measured and predicted C_P/σ versus C_T/σ using SA and SST based transition models, $M_{tip} = 0.69$ (Felker et al. 1985)

A comparison of predicted and measured rotor figure of merit (FM) versus blade loading coefficient (C_T/σ) at a reduced tip Mach number of 0.56 is shown in Fig. 5.8. The computations are performed at four selected blade collective angles of 10°, 12°, 14°, and 16.6° to demonstrate the efficacy of the Stall Delay Method (SDM). The measured data are taken from limited wind tunnel measurements reported by Wadcock and Yamauchi (1998). The CFD predictions without using SDM show a suddenly reduced FM at high collective angles greater than 14°

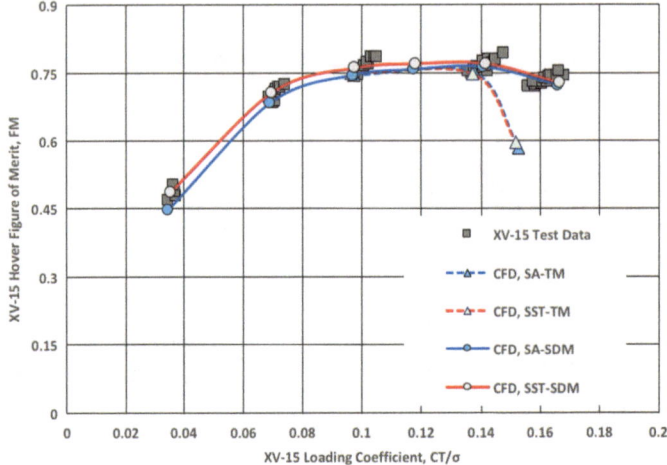

Fig. 5.8 Comparison of measured and predicted FM versus C_T/σ using SA and SST based transition models, $M_{tip} = 0.56$ (Wadcock and Yamauchi 1998)

($C_T/\sigma > 0.14$), departing from the experimental data as shown in Fig. 5.8. However, computations using SDM significantly improve the predictions using both SA and SST-based transition models, especially at a high blade collective angle of 16.6°. Overall, predicted FM using SDM matches well with the measured data at this reduced tip Mach number, indicating that SDM is crucial for an accurate prediction of the rotor performance at high thrust levels or high blade collective angles.

5.1.3   Skin Frictions

Wadcock and Yamauchi (1998), as well as Wadcock et al. (1999) conducted detailed skin friction measurements for the full-scale XV-15 rotor at the NASA Ames 80-ft by 120-ft wind tunnel facility using the oil-film interferometric skin friction technique. The skin friction distributions were measured at six blade radial locations: 0.17R, 0.28R, 0.50R, 0.72R, 0.82R, and 0.94R, where R is the overall blade radius. For the design tip Mach number of 0.69, skin frictions were measured at 3° and 10° blade collective angles. For the reduced tip Mach number of 0.56, skin frictions were measured at 2.9°, 7°, 10°, 14.1°, and 16.6° collective angles. These measured skin friction distributions provide not only the states of laminar and turbulent boundary layers but also flow separations on rotor blade surfaces.

The skin friction values are generally lower in the laminar boundary layer, in contrast to higher values for the turbulent boundary layer. An abrupt increase in skin frictions indicates the onset of transition from laminar to turbulent flow. For helicopter or tiltrotor blades, a common transition mode is a bypass transition

(Chap. 1) caused by the preceding rotor wakes. If the skin friction value is nearly zero before an abrupt increase, this signifies a laminar separation bubble which triggers the so-called separation-induced transition (Chap. 1).

Shown in Fig. 5.9 are comparisons between the predicted and measured skin friction coefficient distributions along the blade radial stations at a low blade collective angle of 3°, where the tip Mach number is 0.69. These skin friction coefficients are calculated based on the blade local speed. At this blade collective angle, the wind tunnel measurements indicate that the transition onset occurs at around 40 % of the blade chord from the leading edge at most blade locations, except for the blade tip (r/R = 0.94) where the transition onset location is shifted to about 60 % of the chord. Bypass transitions are evident since the skin friction values are above zero before the onset of transition. The CFD predictions using both SA-TM and SST-TM have captured the general transition phenomena, but predicted transition onsets are slightly upstream compared with the wind tunnel measurements from the blade middle span up to the tip (r/R = 0.5–0.94), see Fig. 5.9. In addition, the transition onset locations predicted by the SA-TM transition model are slightly earlier than those predicted by SST-TM. This is attributed to different treatments of

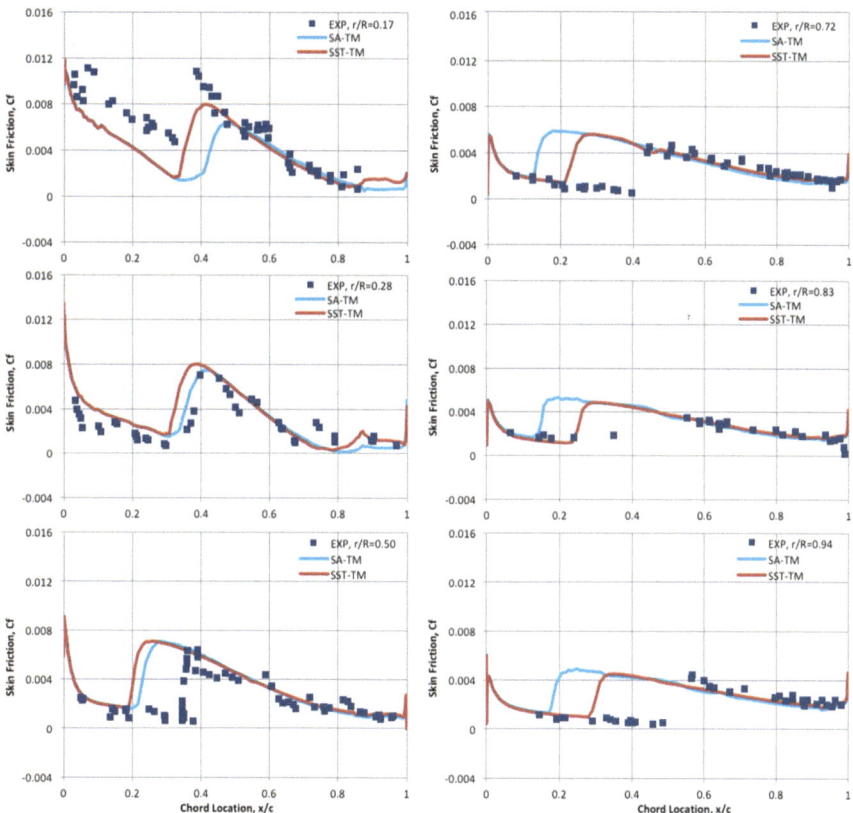

Fig. 5.9 Comparison of measured and predicted skin friction coefficients at θ = 3°, M$_{tip}$ = 0.69

the local free-stream turbulence intensity in the two transition models, because a constant free-stream turbulence intensity is assumed in the SA-TM model while it is calculated locally in the SST-TM model.

At a high blade collective angle of 10°, the measured transition onset locations are shifted towards the blade's leading edge as shown in Fig. 5.10. Predicted transition onset locations are generally matched with the measurements, but slightly earlier transition onsets are still predicted at r/R = 0.72 and 0.83 by the SA-TM model compared to the SST-TM model. Wadcock et al. (1999) observed the leading edge laminar separation bubbles at two blade locations of r/R = 0.5 and 0.75, which trigger the boundary layer transition as indicated by rapid increases of the skin friction values from nearly zero at the onset of transition (Fig. 5.10). The wind tunnel test data also shows the inboard flow separation at the radial locations of r/R = 0.28 and 0.50, where the skin frictions are reduced to nearly zero from the mid-chord to the trailing edge on the upper blade surface (x/c = 0.6–1.0). The inboard flow separations are also captured in the SA-TM and SST-TM based solutions, although a fluctuating pattern of the skin friction coefficients is seen near the blade trailing edge, as shown in Fig. 5.10.

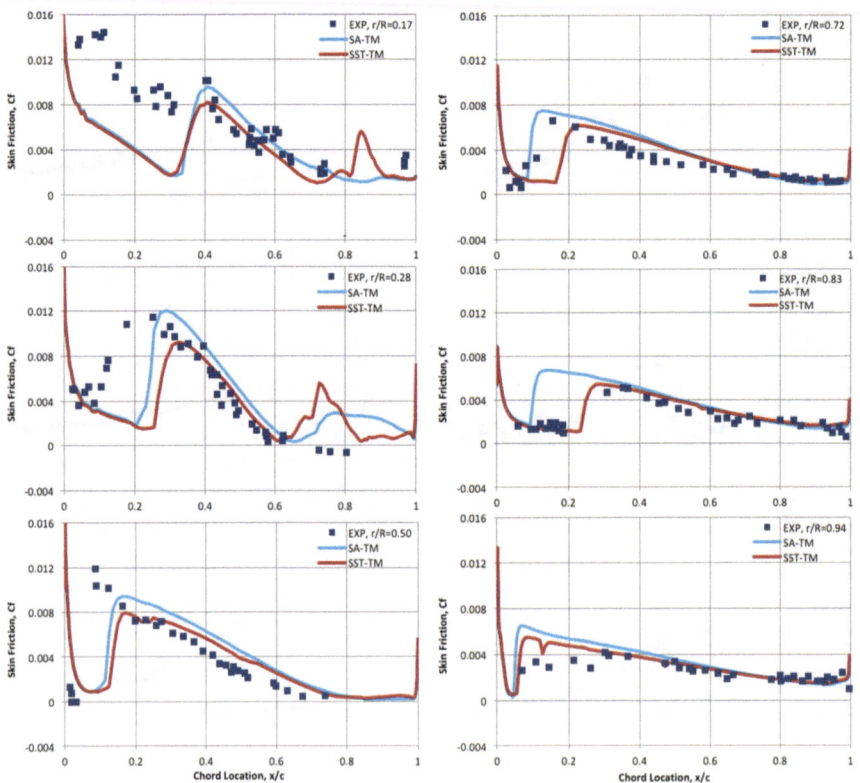

Fig. 5.10 Comparison of measured and predicted skin friction coefficients at $\theta = 10°$, $M_{tip} = 0.69$

Further assessments of the SA-TM and SST-TM transition models are performed at a reduced tip Mach number of 0.56 but at higher blade collective angles. Shown in Fig. 5.11 are comparisons of predicted and measured skin frictions at a collective angle of 14°. The transition onsets triggered by the laminar separation bubbles are captured in the computations from the middle blade span up to the tip (r/R $\geq$ 0.5). The transition onset locations predicted by both SA-TM and SST-TM models become closer because pressure gradients start to dominate the transition process at higher blade collective angles. The mid-chord turbulent flow separations at the inboard radial locations (r/R = 0.28 and 0.50) become more predominant, as indicated by negative or nearly zero skin friction coefficients measured at these locations. Both SA-TM and SST-TM based solutions have captured the mid-chord flow separations at the inboard blade location (r/R = 0.28), but have missed them at the middle span location (r/R = 0.5). Moving outboard (r/R = 0.72, 0.83, 0.94), predicted skin friction distributions by both SA-TM and SST-TM models match very well with the measured values, although some experimental data are missing from the mid-chord to the trailing edge of the blade at radial locations of r/R = 0.83 and 0.94.

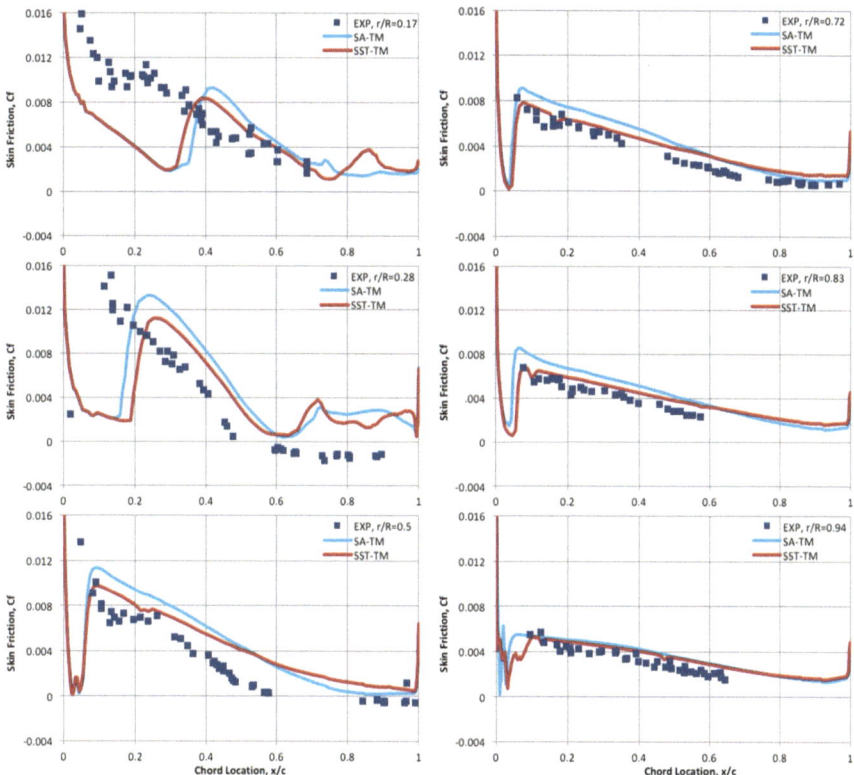

Fig. 5.11 Comparison of measured and predicted skin friction coefficients at θ = 14°, M_{tip} = 0.56

At the highest blade collective angle of 16.6°, the rotor flow is presumed to be
fully turbulent except for the very inboard blade locations of r/R = 0.17 and 0.28,
as illustrated in Fig. 5.12. Even at this high blade collective angle, laminar separation bubbles are still captured in CFD at the blade's leading edge from the middle
blade span up to the tip (r/R ≥ 0.5). There are also extensive reversed flows
observed in the wind tunnel at inboard blade locations of r/R = 0.28 and 0.5, as
shown in Fig. 5.12. Both SA-TM and SST-TM models have captured the inboard
flow separations at r/R = 0.28 but not at r/R = 0.5. This indicates that the size of the
inboard flow separation zone predicted by CFD is smaller than what was observed
in the wind tunnel measurement. No evidence of reversed flows at the outboard
blade stations was reported by Wadcock and Yamauchi (1998).

The above analyses and comparisons about the skin friction distributions indicate that the fundamental aerodynamic features for the XV-15 rotor have been
captured by both SA-TM and SST-TM transition models. These aerodynamic
phenomena include the leading-edge separation bubbles that trigger the boundary
layer transition (separation-induced transition) and flow separations at the inboard
blade locations near the blade's trailing edge due to highly twisted XV-15 rotor

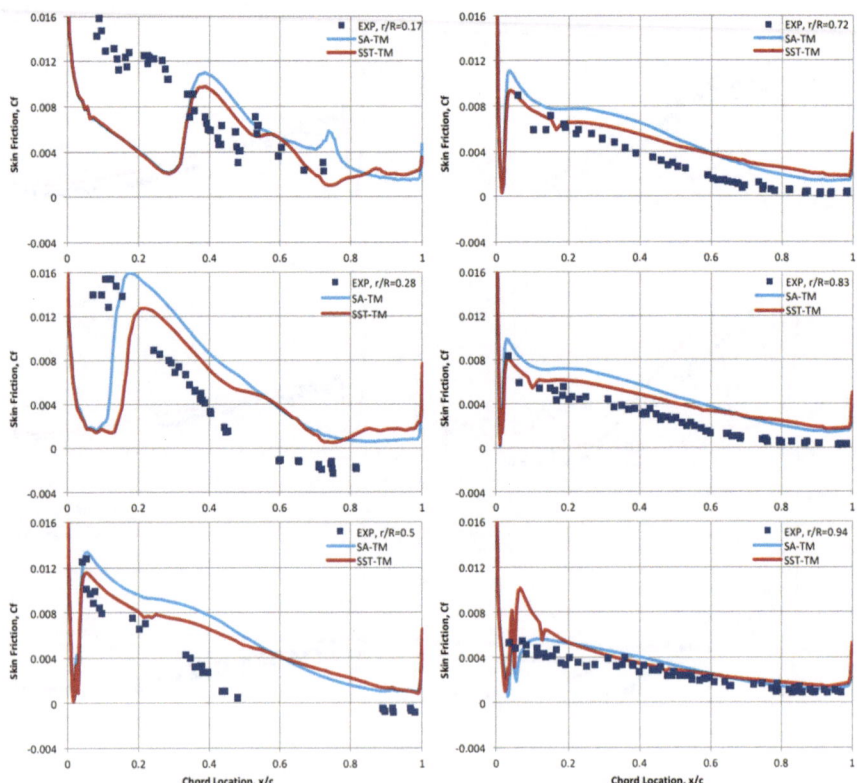

Fig. 5.12 Comparison of measured and predicted skin friction coefficients at θ = 16.6°,
$M_{tip} = 0.56$

blades. In addition, other transition phenomena, such as bypass transitions triggered by trailing tip vortices, are also captured in the computations where the skin friction values are generally higher than zero before the transition occurs.

5.1.4 XV-15 Flow Physics

As discussed above, laminar separation bubbles and inboard flow separations are two predominating flow features observed for the highly twisted XV-15 rotor blades, which are also captured in the CFD computations. However, computational results also revealed an enlarged flow separation predicted in the blade tip region at high thrusts or collective angles, if the Stall Delay Method (SDM) was not used in the simulations. This can be illustrated by the skin friction distributions predicted on the XV-15 rotor at the blade collective angle of 16.6°, which are calculated based on the blade tip speed, and are shown in Figs. 5.13 and 5.14 for the upper and lower surfaces, respectively. An abrupt change in the color scale indicates the onset of transition. A similar transition pattern is seen on the upper surface for the SA-TM and SST-TM models, but discrepancies are seen on the lower surface transition between models, because of different turbulence intensities used in the two models. In addition, the dark colored region in the blade tip indicates an enlarged flow separation predicted by both models without the separation correction, where the skin friction values are

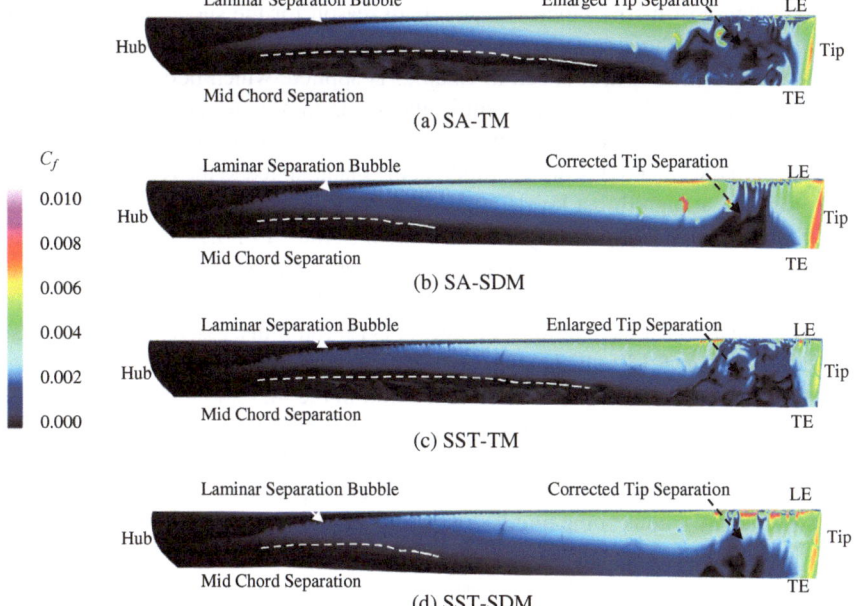

Fig. 5.13 Skin friction coefficient distributions on XV-15 upper surface at $\theta = 16.6°$, $M_{tip} = 0.56$

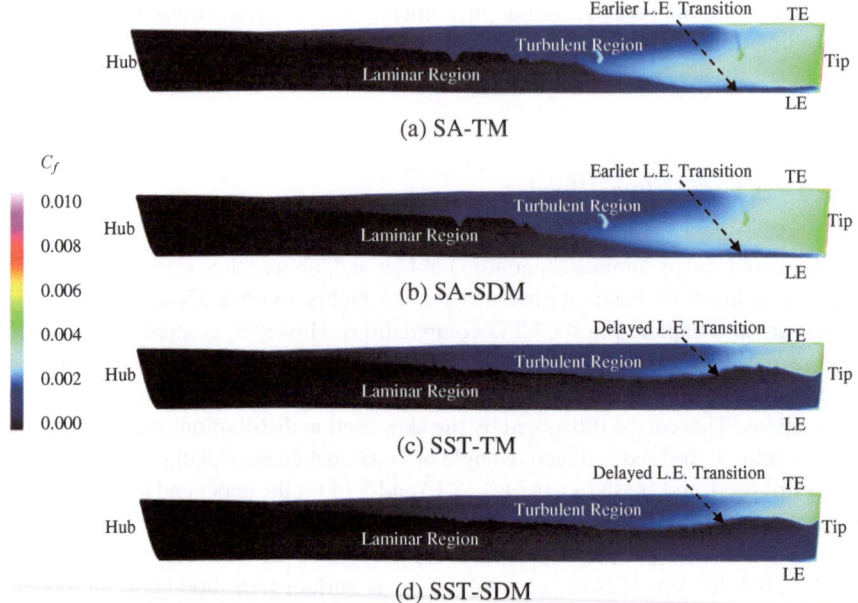

Fig. 5.14 Skin friction coefficient distributions on XV-15 lower surface at $\theta = 16.6°$, $M_{tip} = 0.56$

almost zero as shown in Fig. 5.13a, c. These enlarged tip flow separations have caused an over-prediction of the rotor power consumption and an under-prediction of the rotor figure of merit, which was the main reason for the suddenly reduced FM predicted at high thrusts as shown in Fig. 5.8. To correct this deficiency, the Stall Delay Method (SDM) introduced in Chap. 3 is applied, which boosts the turbulent kinetic energy or eddy viscosity within the separation zone, and thus effectively reduces the enlarged flow separation in the tip region as shown in Fig. 5.13b, d. The correction to the enlarged flow separation at the blade tip led to the recovery of hover FM predicted at high thrust levels, matching it with the measured data as shown in Fig. 5.8. This application clearly demonstrates the efficacy of SDM in preventing the premature flow separations in rotor hover predictions, which is crucial for an accurate prediction of the rotor performance especially at high thrust levels.

The inboard flow separations on the XV-15 rotor discussed earlier are also illustrated in Fig. 5.13, which are marked by white dashed lines. While SDM has effectively reduced the enlarged flow separation in the tip region, it also suppresses the mid-chord flow separation predicted at the inboard blade stations, which are shown in Fig. 5.13b, d for both SA-SDM and SST-SDM models. The reduced inboard flow separation is the reason for the discrepancy between predicted and measured skin frictions at the middle blade span ($r/R = 0.5$) shown in Figs. 5.10 and 5.11. Nevertheless, the benefit of SDM overwhelmingly outweighs the side effect in terms of obtaining an accurate rotor hover prediction and correct rotor flow physics at the high thrust level. Most importantly, both leading edge laminar separation bubbles and inboard flow separations, two key aerodynamic phenomena observed in

Fig. 5.15 Laminar
separation bubble near the
leading edge at outboard
radial station r/R = 0.94,
θ = 16.6°, and M_{tip} = 0.56

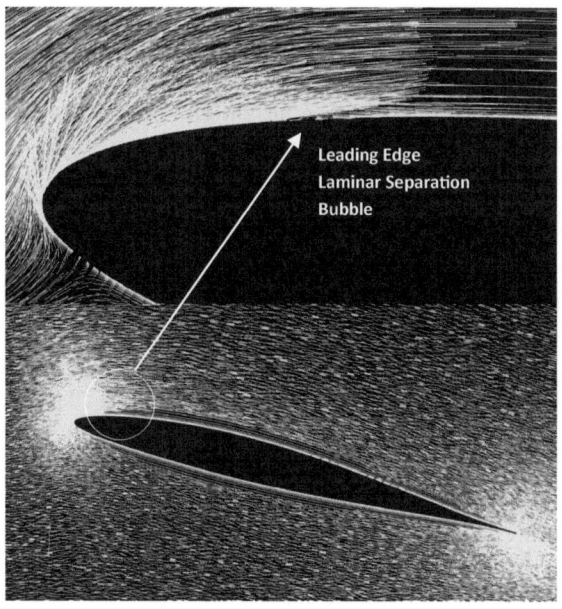

the wind tunnel test (Wadcock and Yamauchi 1998), are successfully captured by
both SA-SDM and SST-SDM transition models using the separation correction
method (SDM). Figure 5.15 shows the leading edge laminar separation bubble
represented by cross-flow velocity vectors at the blade tip of r/R = 0.94. The inboard
flow separation is also illustrated in Fig. 5.16 from the mid-chord to the trailing edge

Fig. 5.16 Mid-chord to
trailing edge flow separation
at inboard radial station
r/R = 0.28, θ = 16.6°, and
M_{tip} = 0.56

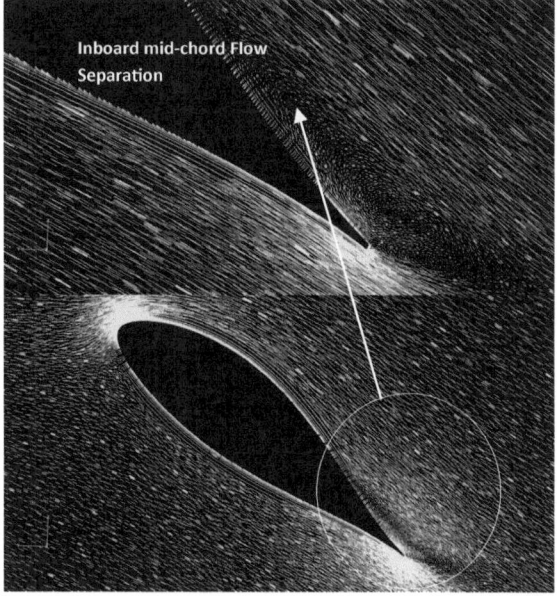

of the blade at an inboard blade location of r/R = 0.28. Successful validations of the current transition models (SA-TM and SST-TM) as well as the Stall Delay Method (SDM) for the XV-15 proprotor should provide a guideline for predicting rotor performance and gaining insight into complex rotor flow physics in the future.

5.2 JVX Proprotor[2]

5.2.1 JVX Geometry and Conditions

The V-22 Osprey was the first military produced tiltrotor aircraft developed by the companies Bell and Boeing. A 0.658-scale V-22 proprotor, the Joint Vertical take-off and landing eXperimental (JVX) rotor shown in Fig. 5.17, was tested at the NASA Ames Research Center Outdoor Aerodynamic Research Facility (OARF) in both hover and airplane mode flight conditions up to an advance ratio of 0.562 (231 knots) (Felker et al. 1987). This Bell-Boeing proprietary rotor system has three blades with a radius of 150 in. The tip chord length is 15.79 in. and the thrust-weighted solidity is 0.1138. Geometric information about the JVX rotor in comparison with the V-22 tiltrotor is listed in Table 5.2 based on the report of Acree (2009).

The JVX rotor was tested at a rotational speed of 576 RPM in hover and around 487–490 RPM in airplane mode. This resulted in a slightly higher tip Mach number of 0.674 in hover and a transonic flow phenomenon (shock wave) in the tip region. The airplane mode was tested for several advance ratios from 0.263 to 0.562. Computational investigations for JVX in hover and airplane mode was performed by Acree (2009) in accordance with these flight test conditions at the sea-level, with an atmospheric temperature of 59 °F and pressure of 2.116×10^4 psf. The flow conditions for JVX in both hover and airplane modes are given in Table 5.3.

Many researchers using various experimental and computational tools have extensively investigated the V-22 aerodynamic performance. Meakin (1995) computed the flow for the V-22 tiltrotor in the presence of a half-span wing using the Navier-Stokes moving overset grid. Brand et al. (2001) measured the hover flow field and download on a 0.658-scale semi-span V-22 model. Potsdam and Strawn (2002) performed Navier-Stokes computations for isolated, half-span, and full-span V-22 tiltrotor in hover configurations. A comparative hover study of V-22 TRAM rotor was recently reported by Chaderjian (2012) using different Navier-Stokes solvers including OVERFLOW (Nichols et al. 2006), FUN3D (Anderson and Bonhaus 1994), and HELIOS (Wissink et al. 2012). He concluded that a good

---

[2]Original work was published in the Proceedings of the American Helicopter Society (AHS) International 70th Annual Forum and Technology Display, 20–22 May 2014, Montreal, Canada.

Fig. 5.17 Bell-Boeing JVX
proprotor tested at NASA
Ames Outdoor Aerodynamic
Research Facility (OARF)
(*Source* http://www.nasa.gov/
centers/ames/images/)

Table 5.2 JVX rotor
geometric characteristics in
comparison with V-22 (Acree
2009)

Parameters	JVX	V-22
Scale referenced to V-22	0.658	1
Rotor radius (in.)	150	228.5
Solidity (thrust weighted)	0.1138	0.105
Tip chord (in.)	15.79	22.0
Taper (tip/root chord)	0.65	0.637

Table 5.3 JVX summary
test and computation
conditions

Parameters	JVX hover	JVX airplane
RPM	576	487, 490
Tip mach	0.674	0.570, 0.575
Tip speed (ft/s)	754	637, 641
Airspeed (knots)	0	100–231
Advance ratio	0	0.263, 0.523

correlation between computed and measured rotor hover FM was achieved by using
the Spalart's Detached Eddy Simulation (DES) (Spalart et al. 2006), combined with
refined surface grid resolution at a spacing of 1 % of the tip chord length in order to
capture the initial tip vortex strength generated by the blade (Chaderjian 2012).

The computational grid topology for the JVX rotor is similar to that used for
XV-15, where a single blade is modeled in a 120° sector of the computational
domain. In addition to properly refined near-body volume meshes, which are shown
in Fig. 5.18, the blade surface resolution is also important for capturing the transition phenomena and obtain an improved prediction of the hover performance
(Sheng et al. 2016b). A maximum surface point spacing of less than 2 % of the tip
chord length is applied to maintain the blade surface resolution for the JVX rotor,
which is shown in Fig. 5.19. Three computational meshes are generated for this

Fig. 5.18 Computational domain with single blade JVX rotor and near-body grid resolution

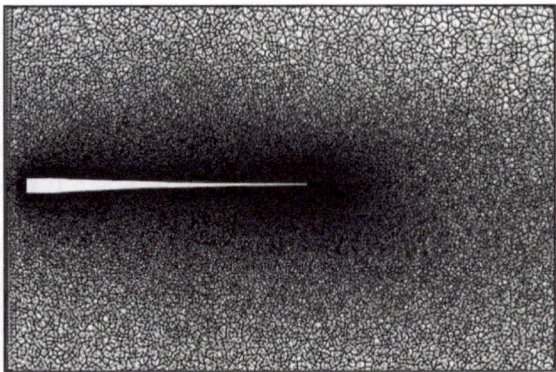

Fig. 5.19 Volume and surface grid resolutions in the tip region of the JVX rotor

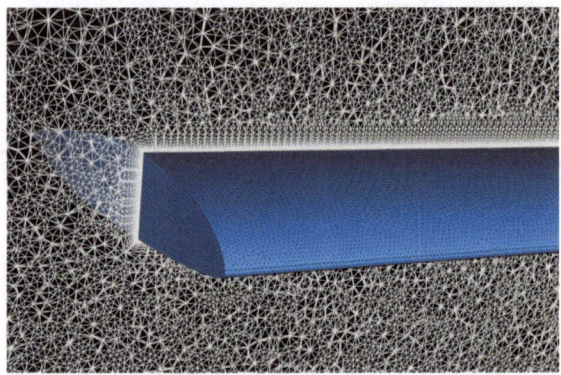

investigation, with a mesh size of 8.9 million nodes for the coarse mesh, 13.7 million nodes for the medium mesh, and 19.3 million nodes for the fine mesh in the single blade computational domain. An axisymmetric boundary condition is applied to periodic surfaces of the computational domain for JVX hover and airplane mode (axial flight) computations.

The JVX rotor performance is investigated in four flow regimes, two collective ranges in hover mode and two advance ratios in airplane mode. In the hover mode, the rotor figure of merit is investigated at blade collective angles ranging from 0° to 16°, where the collective angle of 16° corresponds to the highest thrust level measured in the experiment. The second collective range is from 17° to 22°, which is considered as the high thrust level but no wind tunnel data were measured due to a lack of rotor power. In the airplane mode, the rotor propulsive efficiency is investigated at a lower advance ratio of $\mu = 0.263$ as well as a high advance ratio of $\mu = 0.532$. These flow conditions cover a board flight envelope for the JVX proprotor in both hover and airplane modes. Because each rotor flow field is characterized by different flow physics, suitable numerical techniques are investigated to address specific issues associated with different flow regimes.

5.2.2 Hover Mode

Like the previous XV-15 proprotor, JVX shares many common features such as a highly twisted blade, a low aspect ratio, and a high disc loading. However, an improved hover performance is generally achieved for the JVX rotor compared to the previous XV-15 due to several improvements in its aerodynamic design. The hovering rotor performance is investigated using both S-A and SST based transition models in order to assess their numerical effects on capturing rotor flow physics. Shown in Figs. 5.20 and 5.21 are the predicted figure of merit (FM) and power loading coefficient (C_P/σ) versus the thrust loading coefficient (C_T/σ) obtained on the fine mesh, which are compared with the experimental data at blade collective angles ranging from 0° to 16°. An improved FM is seen for the JVX rotor compared to the XV-15 rotor at the same thrust level. This is due to the enhanced blade profile that eliminates the inboard flow separation experienced by the XV-15 rotor (Fig. 5.3). Predictions using both S-A and SST transition models show agreements with the experimental data, where the maximum difference between the predicted and measured rotor FM is within 2.9 %.

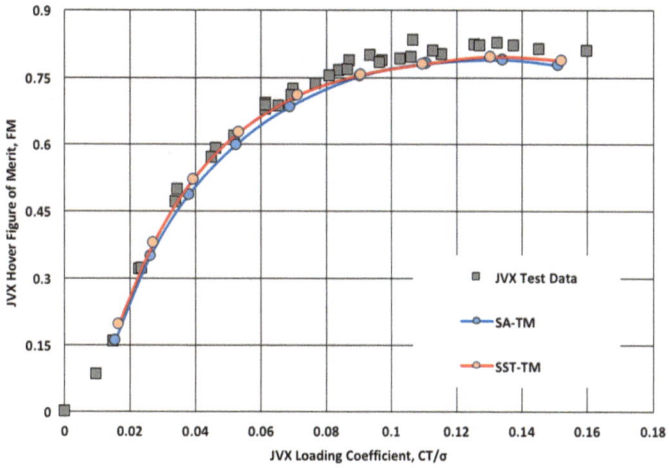

Fig. 5.20 Comparison of predicted and measured FM versus C_T/σ for JVX rotor

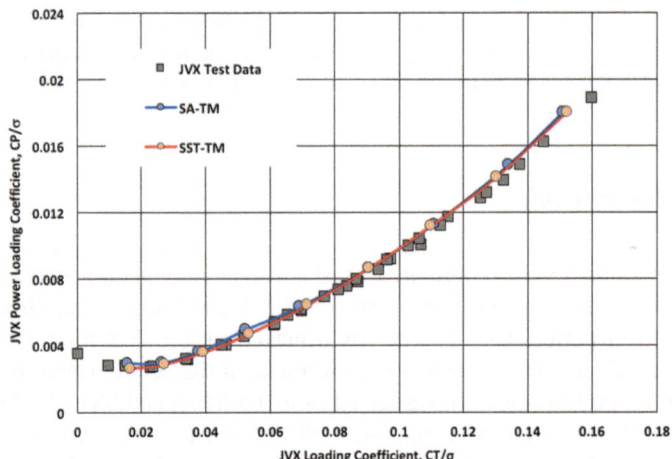

Fig. 5.21 Comparison of predicted and measured C_P/σ versus C_T/σ for JVX rotor

Fig. 5.22 Comparison of intermittency distributions on the JVX upper surface

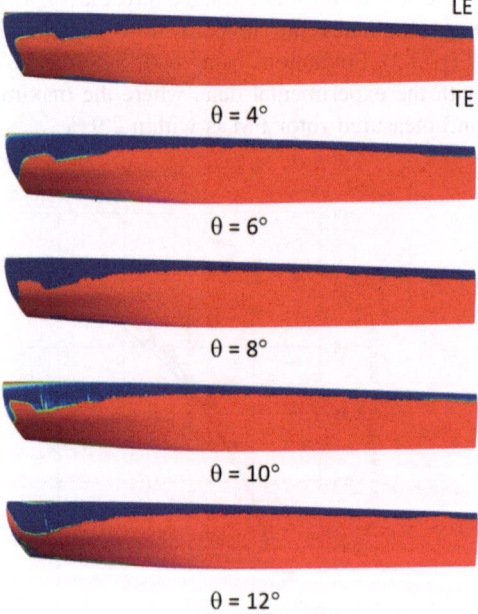

The transition phenomena for the JVX rotor at different blade collective angles can be viewed by intermittency distributions predicted on the upper and lower surfaces, which are shown in Figs. 5.22 and 5.23, respectively. The blue color denotes regions of laminar boundary layers and the red color denotes regions of

Fig. 5.23 Comparison of intermittency distributions on the JVX lower surface

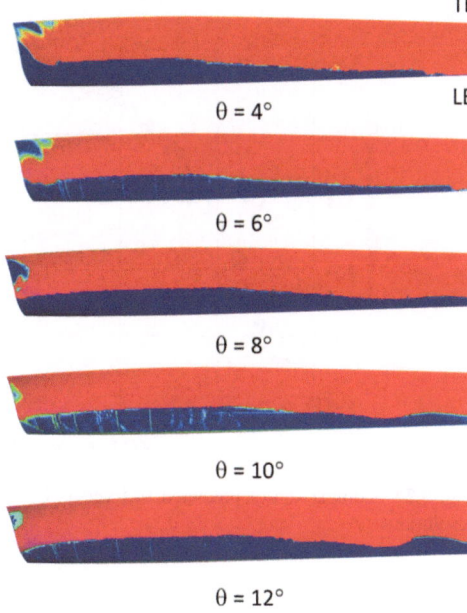

turbulent boundary layers. The laminar boundary layer region is relatively small on both upper and lower surfaces, where the flow turns into fully turbulent quickly leading to an earlier transition onset. It is interesting to note that as the blade collective angle increases, the laminar flow region reduces its size in the blade tip but expands in the blade hub on the upper surface. On the lower surface, however, the trend is opposite. This is due to the highly twisted nature of the JVX proprotor. Although not shown here, both S-A and SST transition models have captured similar transition patterns on the JVX upper surface where strong pressure gradients dominate the transition process. On the lower surface, the bypass transition triggered by the trailing tip vortices is also predicted by both models in the middle region of the blade.

5.2.3   Modelling Issues

Computations for the JVX rotor at blade collective angles less than 16° are performed using both steady and unsteady simulation methods, which yield very close results for the integrated forces and moments of the rotor. Shown in Figs. 5.24 and 5.25 are comparisons of measured and predicted FM and C_P/σ versus C_T/σ using the fully turbulent flow assumption (S-A) and the transition model (SA-TM). The rotor flow field eventually converges to being steady state upon full development of the trailing tip vortices, and predicted forces and moments approach to nearly the

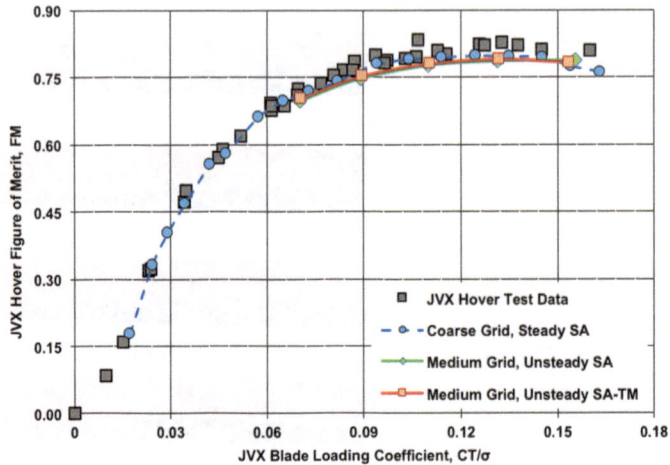

Fig. 5.24 Comparison of predicted FM versus C_T/σ using different turbulence models and grid resolutions

same values in both steady and unsteady runs. The turnaround time is about 15.6 wall-clock hours for a steady solution on the coarse grid, and about 31.2 wall-clock hours for an unsteady solution on the medium grid, using 64 Intel Xeon X5560 cores on a per rotor collective basis.

Grid sensitivity studies are also performed based on the coarse and medium meshes, and computed results are shown in Figs. 5.24 and 5.25. These computed results are similar to those obtained on the fine mesh as shown in Figs. 5.20 and 5.21. Predicted rotor FM and C_P/σ versus C_T/σ curves seem to be less sensitive to the volume mesh resolution than the surface mesh resolution. In fact, excessively refined volume meshes in the off-body region, although helpful for capturing the details of trailing tip vortex structures and rotor wakes, seem to not have a significant impact on the prediction of integrated rotor performance such as thrust and power loadings (Sheng et al. 2016b). However, accurate capturing of the first-passage tip vortex is important to ensure correct inflow conditions for the hovering rotor.

The effect of transition modeling is investigated by comparing the fully turbulent solution (S-A model) with the one obtained with the transition model (SA-TM). These results are also illustrated in Figs. 5.24 and 5.25. It seems that predicted JVX hover performance using the full turbulence model (S-A) yields nearly the same result as using the SA-TM transition model, indicating no significant impact of transition modeling on the JVX hover predictions. Computed skin friction distributions on the JVX upper surface, shown in Fig. 5.26, indicate that a leading edge transition onset is captured by the transition model (SA-TM) but not by the full turbulence model (S-A). In addition, both models have captured similar tip vortex

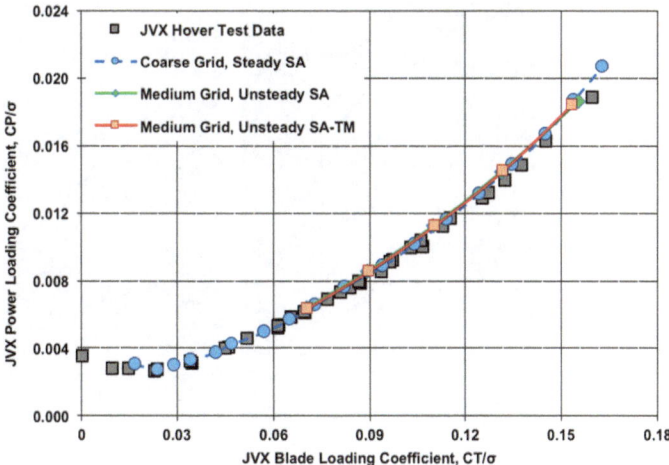

Fig. 5.25 Comparison of predicted C_P/σ versus C_T/σ using different turbulence models and grid resolutions

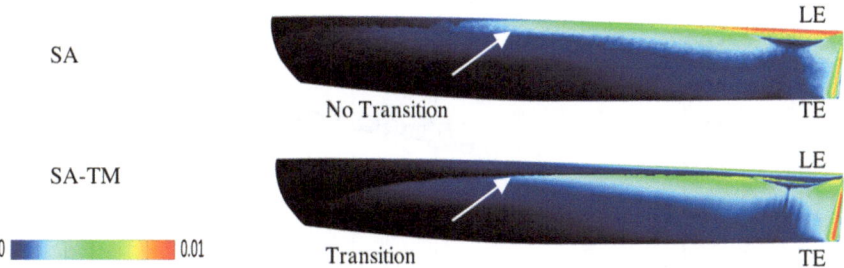

Fig. 5.26 Comparison of predicted skin friction coefficients on the upper surface by SA and SA-TM, $\theta = 16°$

structures as illustrated in Fig. 5.27. These findings are consistent with the previous XV-15 proprotor computations discussed in the previous section.

The ability to accurately predict the hover performance at high thrusts is of practical importance in rotor aerodynamic design and optimization, as rotorcraft engineers constantly push any performance gains in order to expand the flight envelope. To assess the computational capability for the JVX rotor at high thrust levels, computations are performed for the JVX rotor at collective angles ranging from 17° to 22°. Although no experimental data are available for validating the computational results in this collective range, these investigations are informative for gaining understanding of the JVX high thrust performance and associated flow physics, considering satisfactory validations of the current methods obtained for the previous XV-15 predictions. Shown in Figs. 5.28 and 5.29 are the rotor FM and C_P/σ versus C_T/σ predicted by various turbulence modeling methods, which

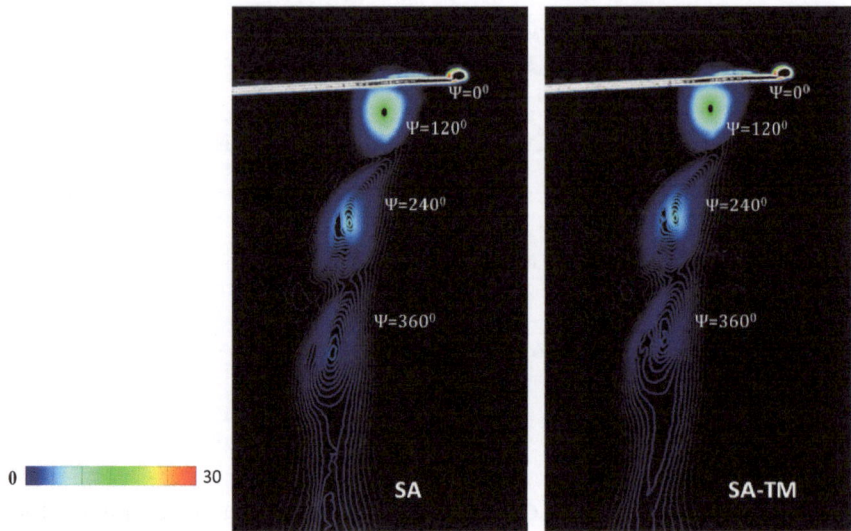

Fig. 5.27 Comparison of predicted tip vortices by SA and SA-TM models, $\theta = 16°$

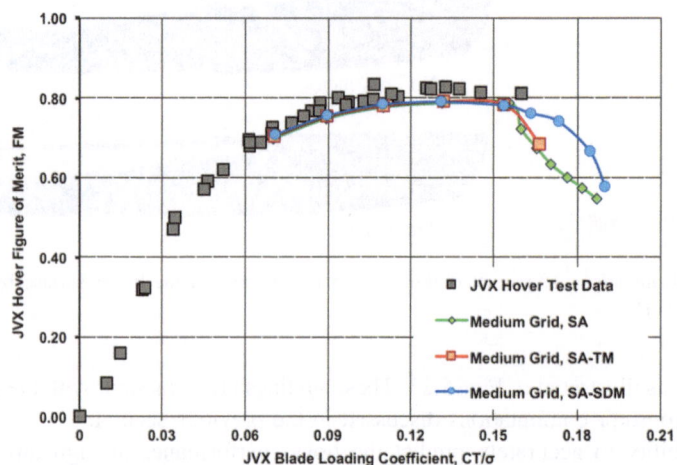

Fig. 5.28 Comparison of predicted FM versus C_T/σ using different turbulence models at high thrusts

indicate a rather distinct hover performance predicted at these high thrust levels. Solutions using both S-A and SA-TM modes (without SDM) show a dramatic decrease in the figure of merit and a sudden increase in the blade power loading coefficient at high blade collective angles greater than 16° (or $C_T/\sigma > 0.16$). However, predictions using SDM produce more reasonable results of the hover

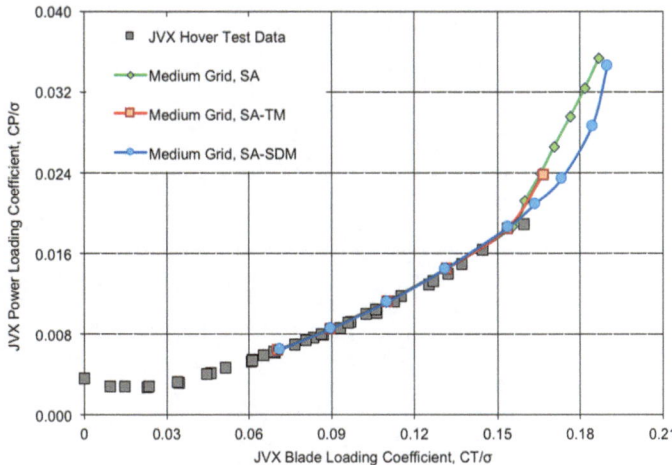

Fig. 5.29 Comparison of predicted C_P/σ versus C_T/σ using different turbulence models at high thrusts

performance, which is consistent with the behavior observed for the XV-15 rotor at high thrust levels ($C_T/\sigma > 0.15$). A closer examination of computed results reveals very different flow patterns predicted in the blade tip region, which are illustrated in Fig. 5.30 as the skin friction distributions. The SA-TM transition model, although successfully capturing the transition phenomenon at the blade leading edge, yields a premature flow separation in the tip region, as does the standard S-A model. Once the flow separates at the blade tip, the rotor inflow condition is severely altered, as shown in Fig. 5.31, which causes a profound impact on the rotor thrust and power characteristics. In other words, incorrect flow physics or premature separation predicted at the blade tip leads to a dramatic reduction of FM at high thrusts, as shown in Fig. 5.28. The Stall Delay Method (SDM), introduced in Chap. 3 for correcting the model's behavior near separation points, has again demonstrated the

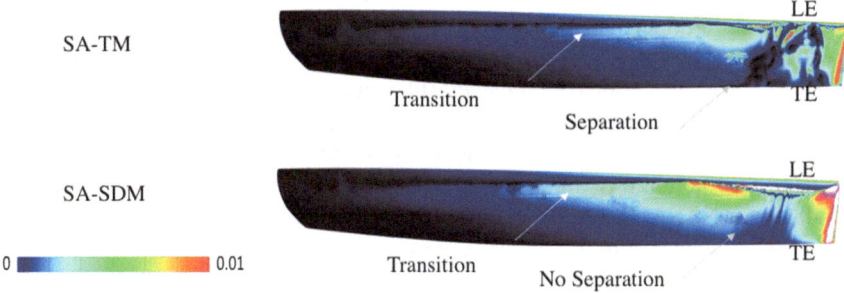

Fig. 5.30 Comparison of predicted skin friction coefficients at high thrust by SA-TM (*upper*) and SA-SDM (*lower*), $\theta = 18°$

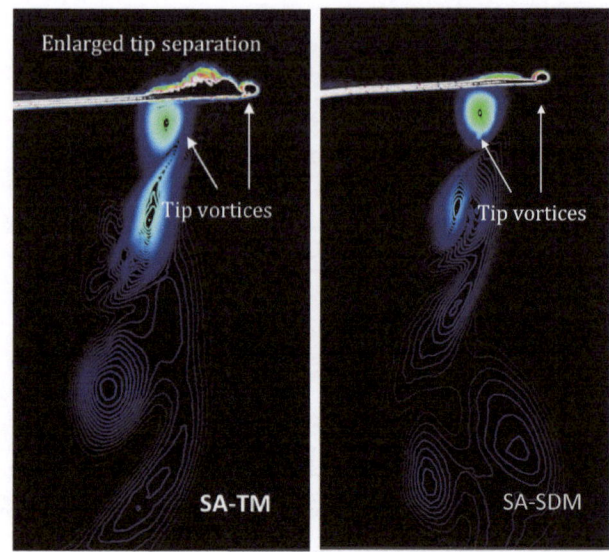

Fig. 5.31 Comparison of predicted tip vortices by SA-TM and SA-SDM models, $\theta = 18°$

ability of eliminating or postponing the premature flow separation encountered by the rotor at high thrusts or blade collective angles, and thus significantly improves the rotor hover predictions. Although there are no experimental data available for validating the computed results for the JVX rotor at this high thrust levels, successful validations for the similar proprotor, XV-15, should provide a reasonable support for the rotor hover performance and associated flow physics predicted for the JVX rotor at high thrusts.

5.2.4 Airplane Mode

One of the key features for the tiltrotor aircraft is its ability to fly at high speeds similar to an airplane. Therefore, computational investigations are carried out for predicting the JVX rotor performance at two airplane modes (axial flow conditions). The JVX airplane propulsive efficiency (PE) and power loading coefficient (C_P/σ) versus the blade loading coefficient (C_T/σ) are computed at two rotor advance ratios: $\mu = 0.263$ and 0.523, where the rotor RPM is around 490 and the tip Mach number is 0.575. These conditions are selected in accordance with the experimental tests conducted at NASA Ames Outdoor Aerodynamics Facilities by Felker et al. (1987).

Computational results of the JVX propulsive efficiency at the lower advance ratio ($\mu = 0.263$) are obtained by changing the blade collective angles from 21° to 30°, and computational results at the higher advance ratio ($\mu = 0.523$) are obtained with the blade collective angles ranging from 36° to 46°. The mesh deformation method used for the JVX hover computations is also used here for performing the blade collective angle sweeps in the airplane mode. The simulation method used for the JVX airplane mode is essentially the same as used in the hover case, where the standard S-A model is used in steady simulations on the coarse mesh, and the SA-TM transition model is used in unsteady simulations on the medium mesh. The computed JVX airplane propulsive efficiency (PE) and power loading coefficient (C_P/σ) versus the blade loading coefficient (C_T/σ) are shown in Figs. 5.32 and 5.33

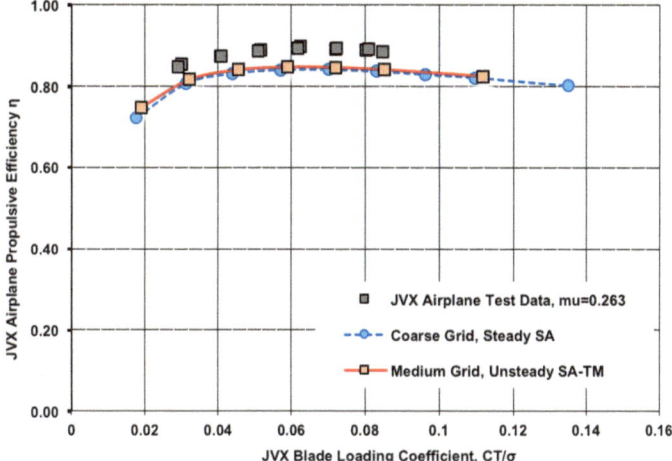

Fig. 5.32 Comparison of predicted and measured propulsive efficiency versus C_T/σ in airplane mode, $\mu = 0.263$

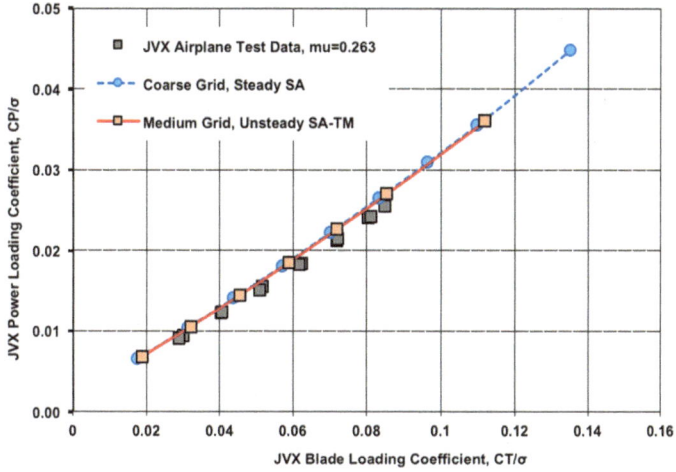

Fig. 5.33 Comparison of predicted and measured C_P/σ versus C_T/σ in airplane mode, $\mu = 0.263$

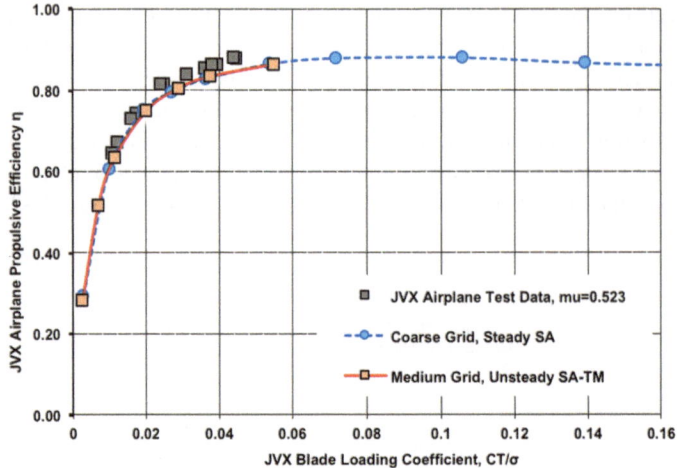

Fig. 5.34 Comparison of predicted and measured propulsive efficiency versus C_T/σ in airplane mode, $\mu = 0.523$

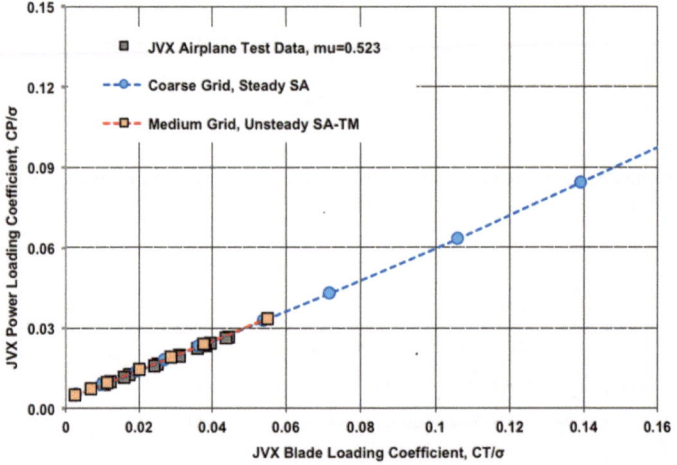

Fig. 5.35 Comparison of predicted and measured C_P/σ versus C_T/σ in airplane mode, $\mu = 0.523$

for the lower advance ratio ($\mu = 0.263$), and in Figs. 5.34 and 5.35 for the higher advance ratio ($\mu = 0.523$). The simulations obtained on both coarse and medium meshes show similar results at each collective range, although the unsteady solutions obtained on the medium mesh are slightly better than those obtained on the coarse mesh using the steady method. In comparison with the experimental data, predicted propulsive efficiencies differ at a maximum of 5.26 % from the measured data at the lower advance ratio ($\mu = 0.263$), and a maximum of 3.36 % from the measured data at the higher advance ratio ($\mu = 0.523$).

Fig. 5.36 Predicted
intermittency distributions,
$\mu = 0.263$, $\theta = 26°$

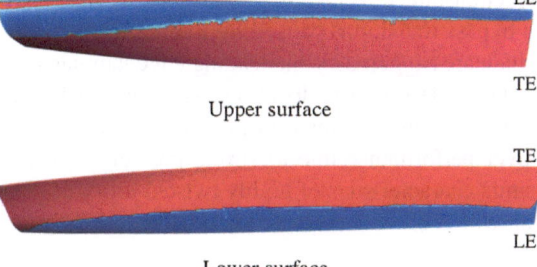

Upper surface

Lower surface

Fig. 5.37 Predicted
intermittency distributions,
$\mu = 0.523$, $\theta = 36°$

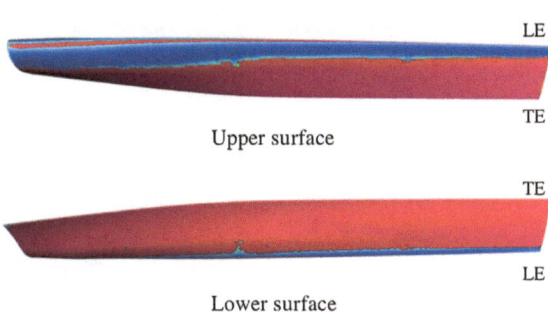

Upper surface

Lower surface

The effect of transition modeling on the JVX airplane performance prediction is also evaluated, which is illustrated in Figs. 5.36 and 5.37 for the intermittency distributions on the upper and lower surfaces, respectively at two advance ratios, respectively. The SA-TM transition model has captured the blade transition in airplane modes at both advance ratios, but no premature flow separation is present at the selected collective angles. Figures 5.36 and 5.37 show enlarged laminar flow regions (blue color) on both upper and lower surfaces compared to the hover case (Figs. 5.22 and 5.23). Increasing the advance ratio results in transition onset locations moving towards the leading edge on the lower rotor surface, but does not change the transition onset location very much on the upper rotor surface due to strong pressure gradients. Again, the transition model does not significantly affect the prediction of the JVX propulsive efficiency in the airplane mode, as similarly discovered in the JVX hover mode.

5.2.5   *JVX Characteristics*

Both JVX and XV-15 are proprotors for tiltrotor aircraft, which are characterized by highly twisted blades, a low aspect ratio, and a high disk loading to provide high speed forward flight capability as well as hover efficiency. These geometric features create complicated three-dimensional aerodynamic phenomena and unique challenges in rotor design and optimization. For example, the JVX rotor is characterized

by a transonic effect leading to a shock wave at the blade tip (Fig. 5.38). In addition
to bypass transitions caused by trailing tip vortices, separation-induced transitions,
which are triggered by the leading edge laminar separation bubbles, are also evident
at the blade outboard locations as shown in Fig. 5.39. For both JVX and XV-15
proprotors, these transition phenomena do not show a large impact on the rotor
hover performance due to strong pressure gradients, which dominate the aerodynamic characteristics of highly twisted tiltrotors. The flow separation, however, can
cause a profound impact on the lift and drag characteristics of the proprotors,

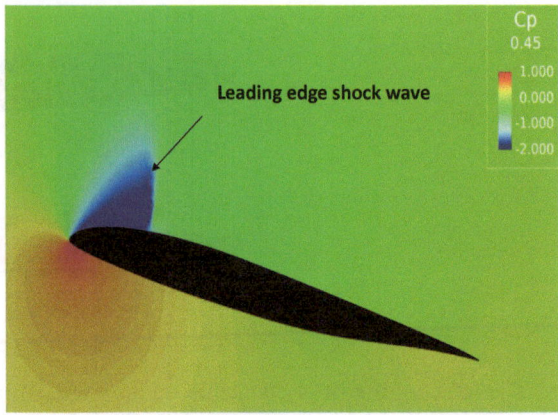

Fig. 5.38 Pressure
coefficient showing the shock
wave at the leading edge of
JVX rotor tip region, $\theta = 18°$

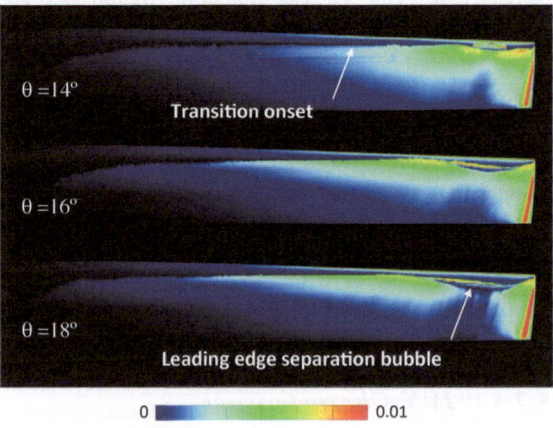

Fig. 5.39 Skin frictions
showing the transition onset
and leading edge separation
bubbles for the JVX rotor,
$\theta = 14°, 16°, 18°$

Fig. 5.40 Laminar
separation bubble near the
leading edge of the JVX rotor
at r/R = 0.96, θ = 18°

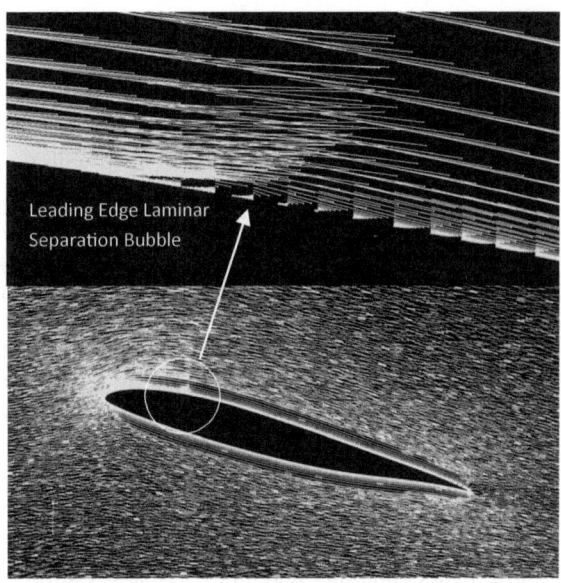

especially in the blade tip region, which affects the rotor figure of merit in hover
flight. A major enhancement for the JVX rotor is the improved blade contour design
that effectively eliminates the inboard flow separations experienced by the XV-15
rotor. Both the experimental investigations (Felker et al. 1987) and the computational simulations (Sheng 2014) have confirmed an improvement in the rotor FM at
high thrust levels for the JVX rotor compared to the XV-15 rotor. While the inboard
flow separation has been largely eliminated for the JVX rotor, the leading edge
separation bubble is still predicted in the CFD computation, especially at a high
blade collective angle as shown in Fig. 5.40. Therefore, proper control of the
separation bubbles without causing massive flow separations or stall at high thrusts
becomes critical to maintain a high aerodynamic efficiency while seeking to expand
the existing flight envelope.

   In summary, both numerics and physics-related modeling aspects are evaluated
in the computations of the JVX rotor in both hover and airplane modes.
Computations have shown a consistent behavior in predicting the aerodynamic
performance and the associated flow physics for the JVX proprotor. The transition
model does not show a significant impact on the prediction of the JVX hover
performance due to the presence of strong pressure gradients. The Stall Delay
Method (SDM), however, has demonstrated a large impact on computations of the
JVX rotor flow field as well as hover performance at high thrust levels or high blade
collective angles.

5.3   S-76 Scaled Rotor[3]

Two advanced technology rotors, a scaled UH-60A Black Hawk rotor and a scaled S-76 rotor, were experimentally investigated by researchers at Sikorsky Aircraft in the mid 80s (Balch and Lombardi 1985a, b) with the goal to identify and quantify attainable improvements in the main rotor hover performance by using advanced geometry rotor tip configurations. Figure 5.41 shows the Sikorsky conventional S-76 helicopter. Unlike the previous proprotors such as the Bell XV-15 (Felker et al. 1985) and the Bell-Boeing JVX (Felker et al. 1987), conventional rotors typically have a high aspect ratio, a longer blade radius, and a moderate blade twist, whose aerodynamic performance is largely impacted by the blade tip shape. The experimental testing suggested that the peak isolated S-76 rotor performance was obtained with a blade tip that combined swept, tapered, and an anhedral profile (Balch and Lombardi 1985a, b).

An accurate prediction of the rotor hover performance is a challenging task. Unlike fixed-wing applications, the rotary blade may encounter trailing tip vortices generated by other blades, laminar to turbulent boundary layer transitions during typical flight conditions, and minor to massive flow separations (stall) at high thrust levels. For conventional rotors such as S-76, aeroelasticity and structural deformation may also affect the aerodynamic performance (Potsdam et al. 2004). Therefore, it is essential that rotor CFD tools should possess not only sufficient numerical accuracy to capture rotor trailing vortices in the near-field but also physics modeling capability to capture critical flow phenomena such as boundary layer transitions and flow separations on the blade surface. While most efforts on capturing hovering rotor performance from first-principles attempted to resolve the rotor wake in the past, relatively less work has been done evaluating the impact of boundary layer transition or separation on conventional rotors. In order to assess the impact of transition modeling on conventional rotor flow field and hover performance predictions, numerical investigations for the conventional S-76 scaled rotor are presented in this section to complement the assessment of S-A and SST transition models on rotor hover performance predictions.

5.3.1   S-76 Geometry and Conditions

The Sikorsky S-76 scaled rotor is a 1/4.71 scaled replica of the actual S-76 aircraft rotor, which possesses $-10°$ linear twist. This four-bladed scaled rotor has a radius of 56.04 in. and a tip chord of 3.1 in. It uses the SC1095 airfoil outboard and SC1094R8 airfoil inboard, which transitions to the SC1394R8 at the root cutout.

---

[3]Original work was published in Journal of Aircraft, Vol. 53, No. 5 (2016), pp. 1549–1560, doi: https://arc.aiaa.org/doi/abs/10.2514/1.C033512.

Fig. 5.41 Sikorsky S-76 conventional helicopter (© Sikorsky Aircraft Corporation 2016. All rights reserved.)

The S-76 rotor hover tests were performed by Balch and Lombardi (1985a, b) at five tip configurations and three tip Mach numbers, which had a test data repeatability of 0.6 % on the figure of merit and 0.4 % on the rotor lift at constant power. Numerical investigations presented in the section include three tip configurations: a 35° swept tapered (60 % tip chord) tip, a rectangular straight tip, and a

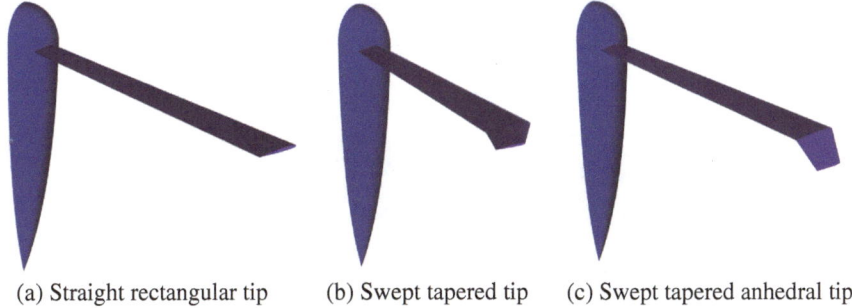

(a) Straight rectangular tip   (b) Swept tapered tip   (c) Swept tapered anhedral tip

Fig. 5.42 S-76 main rotor blade geometry with three different tip shapes

Table 5.4 Summary of S-76 rotor geometric and test conditions

	Straight	Swept tapered	Anhedral
Scale referenced to actual aircraft	0.2123	0.2123	0.2123
Number of blades	4	4	4
Rotor radius (in.)	56.04	56.04	56.04
Solidity (thrust weighted)	0.07043	0.06923	0.06923
Tip chord (in.)	3.1	3.1	3.1
Thrust weighted chord (in.)	3.1	3.047	3.047
Taper (tip/root chord)	1.0	0.60	0.60
Tip Mach number	0.65	0.65	0.65

35° swept tapered (60 % tip chord) with a 20° anhedral tip as shown in Fig. 5.42. The tip Mach number is 0.65. The S-76 blade geometric information and test conditions are given in Table 5.4.

5.3.2 CFD Meshes

To take advantage of the axisymmetric flow field for the isolated S-76 rotor in hover, computations are performed in a single blade computational domain covering a 90° azimuthal sector in the rotating reference frame, which is shown in Fig. 5.43. A vertical cutting plane shows the mesh point clustering around the rotor blade and the center body, and closer views of three tip shapes and surface grid resolutions are shown in Fig. 5.44. Two unstructured meshes, a coarse mesh and a fine mesh, are generated for each tip shape with different surface and volume mesh

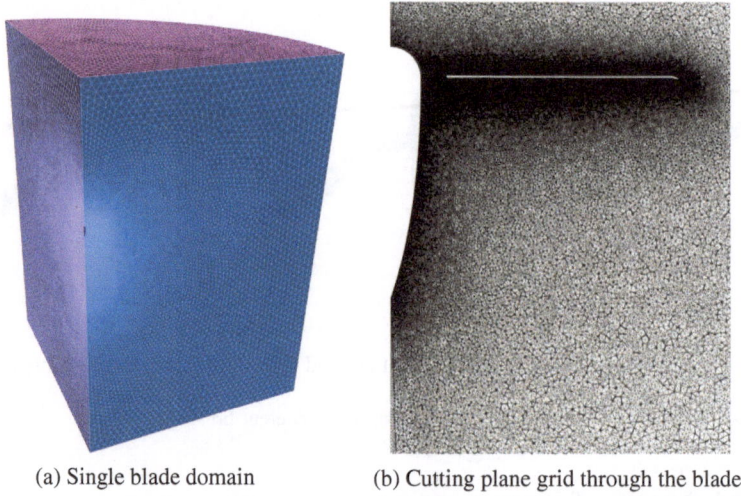

(a) Single blade domain   (b) Cutting plane grid through the blade

Fig. 5.43 S-76 single blade computational domain and grid resolution near rotor and center body

(a) Straight rectangular tip   (b) Swept tapered tip   (c) Swept tapered anhedral tip

Fig. 5.44 S-76 rotor with different tip shapes

Table 5.5 Summary of the S-76 surface point spacing (based on nominal chord length)

Grids	Coarse	Fine
Computational domain	90°	90°
Number of blades	1	1
LE point spacing (0.75R)	0.00484	0.00484
TE point spacing (0.75R)	0.00968	0.00161
Tip point spacing (0.25C)	0.00968	0.00484
Mid span spacing (0.25C)	0.01935	0.01935
Interior domain point spacing	0.16129–0.32258–0.32258	0.09678–0.32258–0.32258

Table 5.6 Summary of the S-76 grid sizes (in million)

	Straight tip		Swept tapered tip		swept tapered anhedral tip	
	Coarse	Fine	Coarse	Fine	Coarse	Fine
Volume nodes	9.5	16.5	9.8	13.9	8.9	13.2
Blade faces	0.7	1.1	0.7	1.0	0.6	0.9
Volume cells	39.3	61.3	40.1	50.5	37.3	51.1

resolutions to investigate the grid sensitivity on rotor hover predictions. The main difference between the coarse and the fine meshes is the surface point spacing in the leading and trailing edges as well as the tip region of the blade. A maximum point spacing of about 1–2 % of the nominal blade chord is adopted for all surface meshes, which is given in Table 5.5. The sizes of the coarse and fine meshes range from 8.8 to 16.5 million volume nodes, 0.6–1.2 million surface faces, and 37.3–61.3 million volume cells for the single S-76 blade. These coarse and fine meshes are equivalent to 35–66 million volume nodes and 149–245 million cells for the four-bladed S-76 rotor system. The mesh size information for the S-76 rotor with three different tip shapes is summarized in Table 5.6.

5.3.3   Effect of Turbulence Models

The S-76 rotor exhibits different flow characteristics at low and high blade collective pitches or thrust levels, which must be addressed correctly in order to obtain an accurate prediction of the rotor performance for the entire blade collective range. To investigate the effect of turbulence models on conventional rotor performance predictions, six turbulence modeling methods are evaluated here: the Spalart-Allmaras turbulence model (SA), the Menter's Shear Stress Transport model (SST), the S-A transition model (SA-TM), the SST transition model (SST-TM), the S-A transition model with the separation correction (SA-SDM), and the SST transition model with the separation correction (SST-SDM). All turbulence modeling methods are evaluated using the same discretization scheme on the fine CFD mesh only. While the experimental data for the S-76 scaled rotor represent only integrated aerodynamic forces in hover, they cannot provide direct validations for the boundary layer transition phenomena predicted on the S-76 rotor. However, numerical validations conducted for the previous XV-15 and JVX proprotors should provide reasonable support for the rotor performance and associated flow physics obtained for the S-76 scaled rotor. The focus in this section is to investigate the effects of different turbulence and transition models in the prediction of a conventional S-76 scaled rotor in comparison with the results obtained for the previous XV-15 and JVX proprotors, in order to gain insight for the flow physics and hover performance associated with conventional helicopter rotors.

Comparisons of predicted FM and C_P/σ versus C_T/σ coefficients for the S-76 scaled rotor with a swept tapered tip are shown in Figs. 5.45 and 5.46, which illustrate the behavior of different models in predicting the S-76 conventional rotor performance. The experimental data are included for assessing the prediction accuracy of the CFD results. Since both S-A and SST turbulence models assume a fully turbulent flow, their predictions consistently show an over-estimate of rotor power consumption and an under-prediction of figure of merit for the entire blade collective range. In contrast, predictions using the SA-TM and SST-TM transition models significantly improve the hover performance (except for 10°) due to their ability to capture the transition phenomenon. This indicates that transition modelling has a greater impact on the S-76 rotor prediction than on the previous two proprotors. Because the S-76 conventional rotor has a high aspect ratio and a moderate blade twist, the pressure gradients over the blade surfaces are not as strong as in those proprotors. Therefore, the laminar boundary layer and its viscous

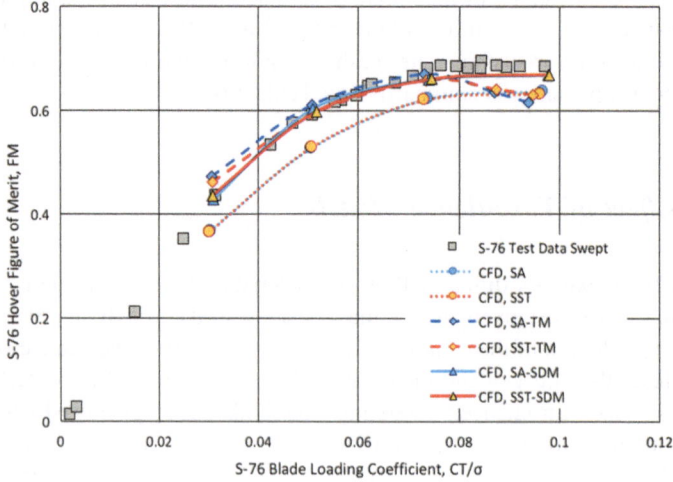

Fig. 5.45 Comparison of predicted and measured FM versus C_T/σ for S-76 rotor with swept tapered tip using different models

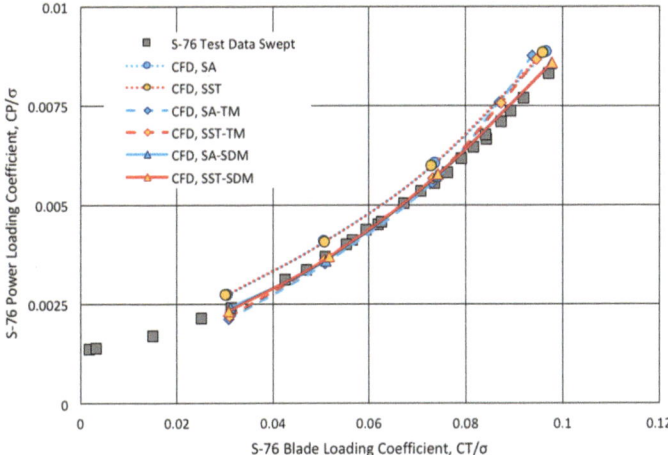

Fig. 5.46 Comparison of predicted and measured C_P/σ versus C_T/σ for S-76 rotor with swept tapered tip using different models

shear stress can significantly affect for the integrated aerodynamic force and power for conventional rotors.

While both S-A and SST based transition models have produced a marked improvement for the S-76 hover prediction, noticeable discrepancies still exist between the predicted and measured FM at higher collective angles or thrust levels. As shown in Figs. 5.45 and 5.46, the CFD predictions show a quickly reduced FM at collective angles greater than $10°$ (or $C_T/\sigma > 0.08$), departing from the measured value at this thrust level. Numerical experiments indicated that this behavior could not be corrected by simply increasing the surface or volume mesh resolutions, or using higher order spatial discretization schemes. In fact, this behavior of suddenly reduced FM predicted at high blade collective angles was also reported by several other researchers in the recent S-76 rotor hover predictions hosted at the American Institute of Aeronautics and Astronautics (AIAA) Applied Aerodynamics Conferences (Jung et al. 2014; Hwang et al. 2015; Min and Wake 2016; Gardarein and Le Pape 2016). An example of the S-76 hover prediction performed by Gardarein and Le Pape (2016) is shown in Figs. 5.47 and 5.48, where the ONERA structured grid CFD code *elsA* (http://elsa.onera.fr/elsA/doc/refdoc.html) was used in the computation. Their results indicate a sudden drop of FM for the S-76 rotor at blade collective angles greater than $10°$ using both Wilcox and Kok k-ω full turbulence model and Menter-Langtry $\gamma - \tilde{Re}_{\theta t}$ transition model (Minot et al. 2015).

The analysis of the flow field for the S-76 conventional rotor at high thrusts reveals enlarged flow separations predicted in the tip region, similar to the behavior found for the previous two proprotors (XV-15 and JVX) at high thrust levels. Shown in Figs. 5.49 and 5.50 are the skin friction distributions predicted by both S-A and SST based models at a $12°$ collective angle, respectively. These skin friction coefficients are calculated based on the blade tip speed. Predictions without using the

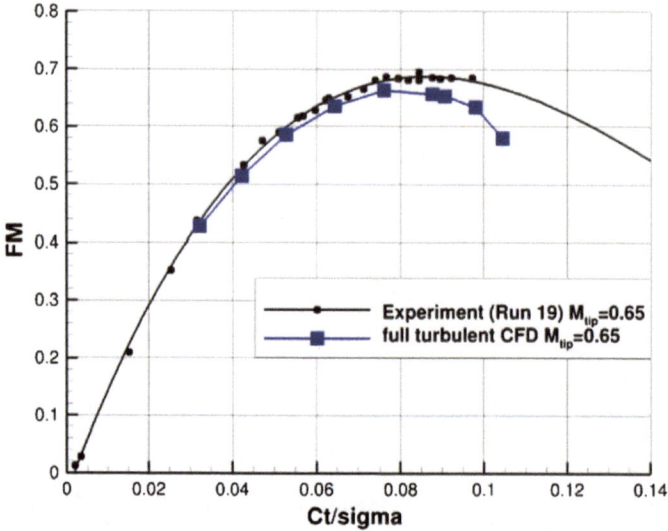

Fig. 5.47 Comparison of predicted and measured FM versus C_T/σ for S-76 rotor with swept tapered tip using ONERA elsA code (courtesy of Gardarein and Le Pape 2016)

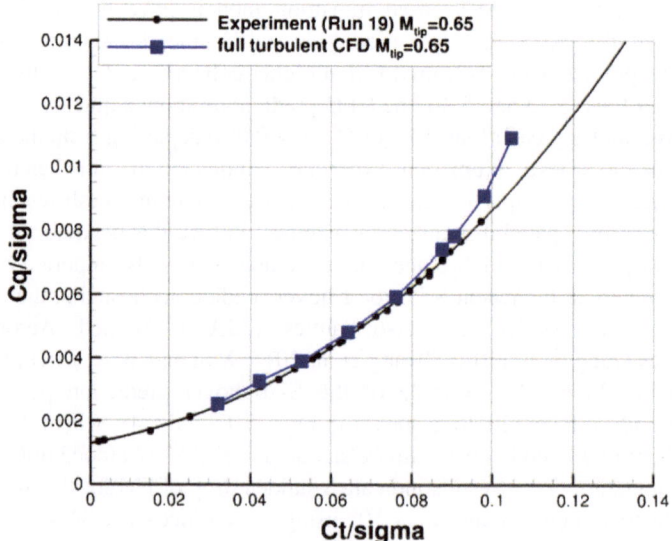

Fig. 5.48 Comparison of predicted and measured C_Q/σ versus C_T/σ for S-76 rotor with swept tapered tip using ONERA elsA code (courtesy of Gardarein and Le Pape 2016)

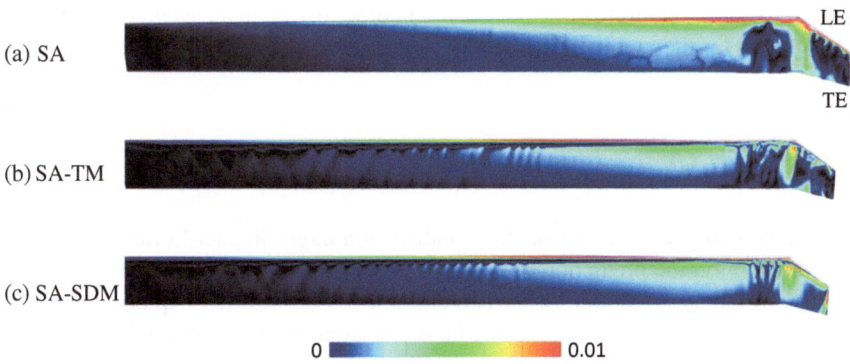

Fig. 5.49 Skin friction coefficients predicted by different SA based models, $\theta = 12°$

separation correction (SDM) (Figs. 5.49a, b and 5.50a, b) show signs of overly predicted flow separations in the tip region. This is the main cause of an increase in rotor power consumption and a sudden reduction in figure of merit as shown in Figs. 5.45 and 5.46. In addition, other researchers (Jung et al. 2014; Gardarein and Pape 2016; Min and Wake 2016) also reported an over prediction of flow separations in the S-76 rotor tip region that caused the sudden drop of figure of merit at high collective angles. All these computational results indicate a common issue faced by the RANS modeling approach, which is the over prediction of separated flows due to an under-estimate of turbulence close to separation points (Aupoix et al. 2011). In order to correct this numerical deficiency, the Stall Delay Method (SDM) described earlier is applied to correct the turbulence models' behavior near separations. As shown in Figs. 5.49c and 5.50c, the premature flow separation in the blade tip region is successfully suppressed using SDM and the rotor hover performance is thus recovered and matched with the experimental data (Figs. 5.45 and 5.46). These numerical results indicate again that the correction to the turbulence model's

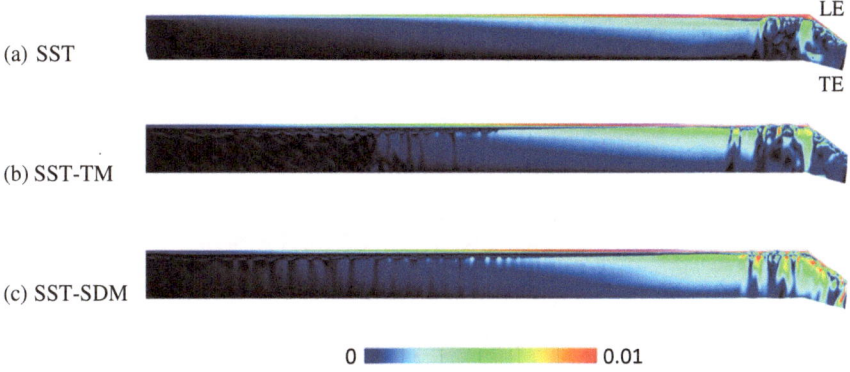

Fig. 5.50 Skin friction coefficients predicted by different SST based models, $\theta = 12°$

behavior near separation points is essential in order to obtain an accurate rotor hover prediction over the entire collective range or thrust level.

5.3.4   S-76 Characteristics

The above computational results have indicated a larger impact from the transition phenomenon on the prediction of the S-76 hover performance than on the previous two proprotors (XV-15 and JVX). Therefore, it is worth further exploring this flow feature on the S-76 rotor in order to gain understanding of its influence on the rotor performance. Predicted intermittency distributions on the upper and lower surfaces of the S-76 rotor at various collective angles are shown in Figs. 5.51 and 5.52, respectively. In the figures, the blue color denotes regions of laminar boundary layers and the red color denotes regions of turbulent boundary layers. It is seen that there is a sizable laminar flow region developed on the S-76 rotor surfaces due to conventional helicopter blade design. The laminar boundary layer covers a large portion on the upper surface for the blade collective angle up to 8°, as shown in Fig. 5.51. On the lower surface, the majority of the blade surface is covered by laminar flows as shown in Fig. 5.52. The bypass transition triggered by trailing tip vortices is evident at high blade collective angles such as 12°. Because of sizable laminar flows developed on the S-76 rotor surfaces, the effect of laminar boundary layer should be correctly captured for an accurate estimate of the rotor lift and drag characteristics. Therefore, the ability to capture the laminar boundary layer and associated transition phenomena becomes important in the prediction of rotor hover performance for conventional rotors. Although no experimental data are available

Fig. 5.51 Intermittency distributions on the S-76 upper surface (*blue* laminar, *red* turbulent)

to provide direct validations of the transitional flow phenomena predicted here, reasonable confidence should be established based on the successful validation of the current transition models in the previous full-scale XV-15 proprotor.

In the previous investigations for the XV-15 and JVX proprotors, the leading edge laminar separation bubbles were observed at high blade collective angles. These separation bubbles trigger the boundary layer transition on the blade surface, called the separation-induced transition, which is a predominating phenomenon for these proprotors especially for XV-15. In the S-76 scaled rotor, the leading edge

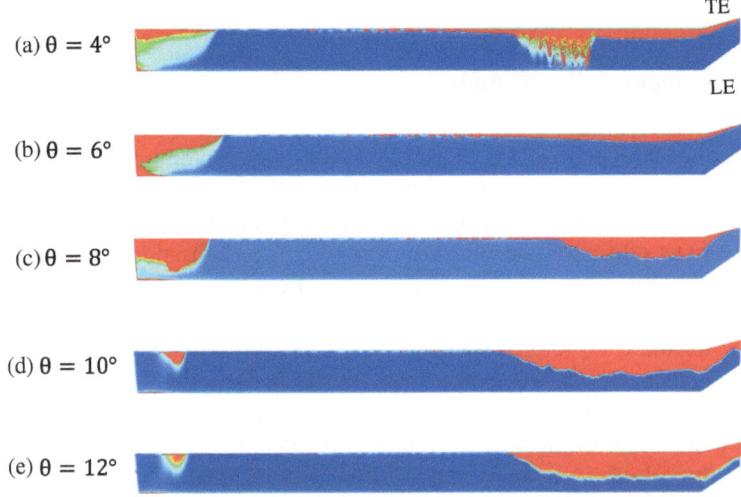

Fig. 5.52 Intermittency distributions on the S-76 lower surface (*blue* laminar, *red* turbulent)

Fig. 5.53 Laminar separation bubble near leading edge at radial station r/R = 0.50, θ = 12°

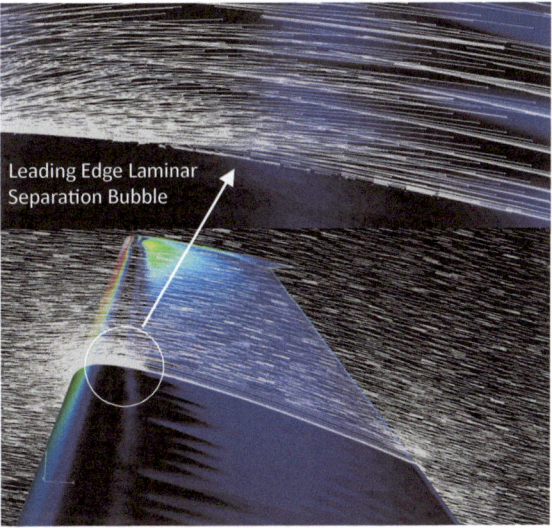

separation bubbles are also predicted on the upper surface at most blade collective angles from the inboard blade location up to the blade tip. At a blade collective angle of 4°, however, a leading edge separation bubble is formed on the lower surface to cause an irregular transition pattern on the S-76 rotor, as shown in Fig. 5.52. Figure 5.53 shows the cross-flow velocity vectors at a mid-span blade location (r/R = 0.5), where a leading edge separation bubble is clearly visible. However, due to a relatively flat blade profile of the S-76 rotor that is very different from highly twisted proprotors, no inboard flow separation is predicted on the S-76 rotor, although it has been observed and predicted for the XV-15 proprotor.

5.3.5 Effect of Tip Shapes

The rotor blade tip plays an important role in the overall rotor performance, because the highest pressure loading, the highest blade Mach number, and the strongest interaction with tip vortices all occur at the blade tip. Different blade tip shapes create various effects on the pressure distributions and the tip shedding vortices, which yield disparities in rotor performance. Both the experiments (Balch and Lombardi 1985a, b) and the computations indicate a similar range of working margins for the S-76 rotor at the same tip Mach number, but a slightly different rotor FM created by different blade tips. A determining factor for the rotor FM calculation is the accurate prediction of power consumption required to generate the same amount of thrust. In order to accurately predict the rotor lift and power values, all rotor flow physics, including the first blade-generated tip vortex, blade transition

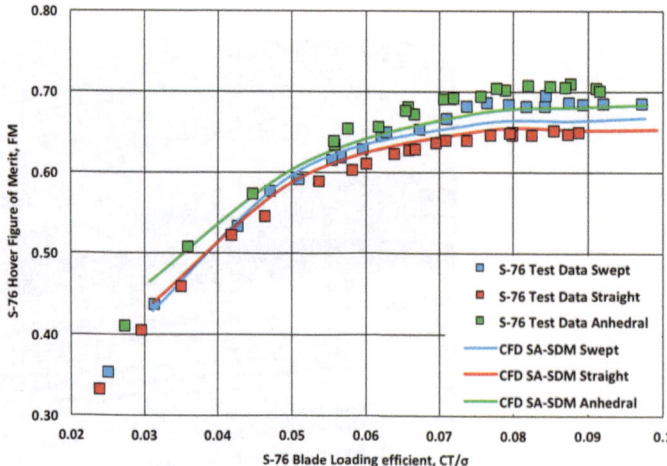

Fig. 5.54 Comparison of predicted and measured FM versus C_T/σ for three tip shapes, SA-SDM model

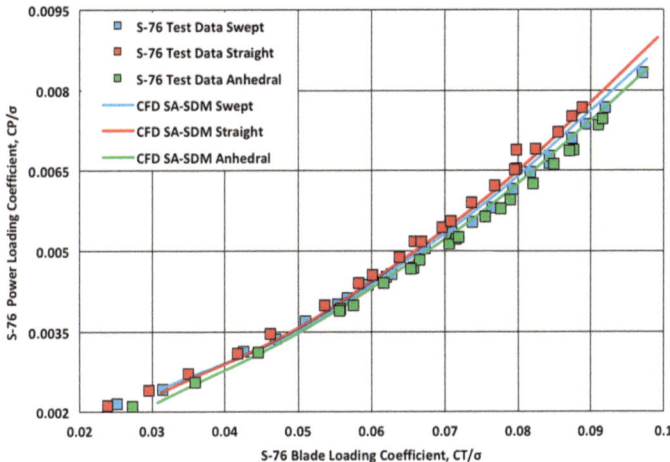

Fig. 5.55 Comparison of predicted and measured C_P/σ versus C_T/σ for three tip shapes, SA-SDM model

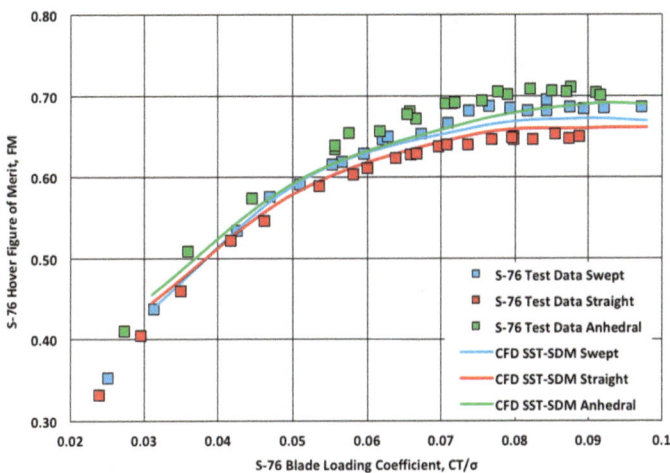

Fig. 5.56 Comparison of predicted and measured FM versus C_T/σ for three tip shapes, SST-SDM model

phenomenon, and flow separation at high collective pitch angle must be captured correctly.

Effects of different tip shapes on the S-76 rotor performance are shown in Figs. 5.54, 5.55, 5.56 and 5.57. The experimental and numerical results both show that the swept tapered rotor tip with anhedral needs the least power, the swept tapered one needs a little more power, and the straight tip design needs the most power throughout the thrust range, as shown in Figs. 5.55, 5.56 and 5.57. Likewise,

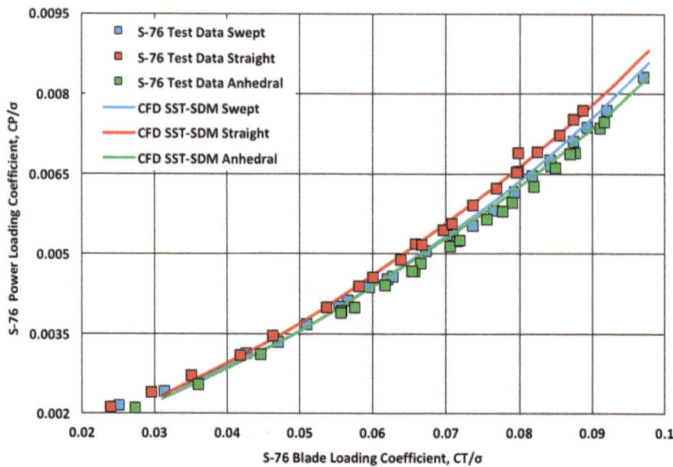

Fig. 5.57 Comparison of predicted and measured C_P/σ versus C_T/σ for three tip shapes, SST-SDM model

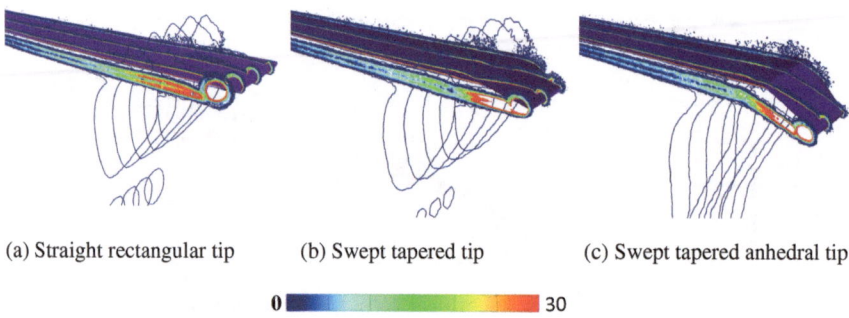

(a) Straight rectangular tip   (b) Swept tapered tip   (c) Swept tapered anhedral tip

Fig. 5.58 Tip vortices generated by different S-76 blade tip shapes

the swept tip with anhedral has the best FM, the swept tapered tip is in the middle, and the straight tip has the lowest efficiency as shown in Figs. 5.54 and 5.56. In general, a small amount of taper in the tip can improve the rotor FM in hover, however, too much taper may lose this benefit due to the higher profile drag at the small tip chord Reynolds numbers (Leishman 2006). The swept tip design reduces the Mach number normal to the leading edge, and thus increases the local rotor advanced ratio in hover. In addition, the swept tip design can affect the formation, location, and shape of the blade tip vortex, which is shown in Fig. 5.58. Comparative performance analysis indicates that swept tapered tips with and without anhedral apparently improve the rotor performance at high thrust levels (C_T/σ from 0.06 to 0.1) compared to the straight tip, and a minor advantage with anhedral compared to the one without anhedral. No marked improvement is seen in the swept tapered tip compared to the straight tip at the low thrust level (C_T/σ from

0.028 to 0.046). This indicates that the swept tapered tip alleviates the tip vortex effect in the tip region at the high thrust level. The combination of the swept and tapered tip designs improves the rotor performance significantly at the high thrust level, and introducing a 20° anhedral to the swept tapered tip further increases the rotor performance at low thrust level. The computational results performed using both S-A and SST transition models have captured these performance features for different rotor tips, and have showed excellent agreements with the experimental data.

5.4   Summary

In this chapter, three realistic tiltrotor and helicopter rotor blades are investigated using the Langtry-Menter $\gamma - \widetilde{Re}_{\theta t}$ transition model, coupled with the Spalart-Allmarsa (S-A) one-equation and Menter's SST two-equation turbulence models. The rotor hover performance and associated flow field are evaluated in detail for each rotor. Numerical results showed that both S-A and SST transition models are capable of predicting the boundary layer transition phenomena on three-dimensional realistic rotors with acceptable accuracy. The Stall Delay Method (SDM) introduced in this book is demonstrated to prevent numerically induced premature flow separation in the tip region at high thrust levels. These modeling capabilities are essential to achieve an accurate prediction of the rotor performance over the entire rotor collective range. Computational investigations for both conventional rotor and tiltrotor blades help interested readers gain improved understanding of complicated rotor flow phenomena, such as leading edge separation bubbles and inboard flow separations. These findings, along with computational modeling techniques presented in this book, may provide a practical guideline and insight for aerodynamic design and optimization of rotorcraft in the future. Concluding remarks are provided regarding computational investigations for the S-76 conventional rotor and XV-15 and JVX proprotors:

1. Both conventional and tiltrotor blades are characterized by trailing tip vortices and leading edge separation bubbles, which are predominant at high thrust levels. The trailing tip vortices cause the so-called bypass transition, while the leading edge separation bubbles trigger the so-called separation-induced transition over the blade surfaces. These aerodynamic phenomena show different level of influences on the thrust and power characteristics of the hovering rotors.
2. Transition phenomenon has a larger impact on the aerodynamic performance for conventional rotor than for tiltrotor blades. This is because of significant laminar flows developed on conventional rotors due to relatively high aspect ratios and moderate blade twists. For highly twisted proprotors, however, turbulent flows are predominant over the blade surfaces because of strong pressure gradients that promote earlier boundary layer transition from laminar to turbulent flows.

3. The Langtry-Menter $\gamma - \widetilde{Re}_{\theta t}$ transition model, coupled with the Spalart-Allmaras (S-A) turbulence model and Menter's Shear Stress Transport (SST) turbulence model, has demonstrated the ability to capture the transition phenomena associated with both conventional rotor and tiltrotor blades. The SA-TM and SST-TM models generally show similar transition patterns predicted on the upper rotor surface, but slight different patterns on the lower rotor surface due to different treatments of the local turbulence intensity in the two models.

4. The new separation correction, or Stall Delay Method (SDM), introduced in this book has shown the ability to correct enlarged flow separations in the blade tip region at high blade collective angles, and thus has significantly improved the hover performance prediction for both conventional rotors and tiltrotors at high thrust levels. The introduction of SDM into the $\gamma - \widetilde{Re}_{\theta t}$ transition model provides a means to obtain an accurate prediction of the rotor hover performance over the entire blade collective angles or thrust levels.

References

Acree CW (2009) JVX proprotor performance calculations and comparisons with hover and airplane-mode test data. In: NASA/TM-2009-215380, Ames Research Center, Moffett Field, California, April 2009

Anderson WK, Bonhaus DL (1994) An implicit upwind algorithm for computing turbulent flows on unstructured grids. Comput Fluids 23:1–21

Aupoix B, Arnal D, Bezard H et al (2011) Transition and turbulence modeling. J AerospaceLab (2), March 2011

Balch DT, Lombardi J (1985a) Experimental study of main rotor tip geometry and tail rotor interactions in hover. Vol. I—text and figures. In: NASA CR 177336 vol 1, Feb 1985

Balch DT, Lombardi J (1985b) Experimental study of main rotor tip geometry and tail rotor interactions in hover. Vol. II—run log and tabulated data. In: NASA CR 177336 vol 2, Feb 1985

Brand AG, Peryea MA, Wood TL et al (2001) Flowfield and download measurements and computation of a tiltrotor in hover. In: Proceedings of the AHS 57th annual forum, Washington, DC, 9–11 May 2001

Chaderjian NM (2012) Advances in rotor performance and turbulent wake simulation using DES and adaptive mesh refinement. Presented at 7th international conference on computational fluid dynamics, Big Island, Hawaii, 9–13 July 2012

Felker FF, Betzina MD, Signor DB (1985) Performance and loads data from a hover test of a full-scale XV-15 rotor. In: NASA TM-86833

Felker FF, Signor DB, Young LA et al (1987) Performance and loads data from a hover test of a 0.658-scale V-22 rotor and wing. In: NASA TM-89419, April 1987

Gardarein P, Le Pape A (2016) Numerical simulation of hovering S-76 helicopter rotor including far-field analysis. AIAA-2016-00034. Presented at 54th AIAA aerospace sciences meeting, AIAA SciTech conferences, 4–8 Jan 2016, Sab Diego, California

Hwang JY, Choi JH, Kwon OJ (2015) Assessment of S-76 rotor aerodynamic performance in hover on unstructured mixed meshes. AIAA 2015 SciTech conferences, 5–9 Jan 2015, Kissimmee, Florida

Jung MK, Hwang JY, Kwon OJ (2014) Assessment of rotor aerodynamic performances in hover using an unstructured mixed mesh method. Paper presented at the 52nd AIAA aerospace sciences meeting, AIAA SciTech, 13–17 Jan 2014

Kaul UK (2012) Effect of inflow boundary conditions on hovering tilt-rotor flows. In: 7th international conference on computational fluid dynamics (ICCFD7), Big Island, Hawaii, 9–13 July 2012

Kaul UK, Ahmad J (2011) Skin friction predictions over a hovering tilt-rotor blade using OVERFLOW2. AIAA Paper 2011-3186, 29th AIAA applied aerodynamics conference, 27–30 June 2011, Honolulu, Hawaii

Leishman JG (2006) Principles of helicopter aerodynamics, 2nd edn. Cambridge Aerospace Series, Cambridge University Press, Cambridge, pp 292–295

Maisel MD, Giulianetti DJ, Dugan DC (2000) The history of the XV-15 tilt rotor research aircraft. In: NASA SP-2000-4517

Meakin RL (1995) Unsteady simulation of the viscous flow about a V-22 rotor and wing in hover. AIAA-95-3463. Paper presented at the 20th AIAA atmospheric flight mechanics conference, Baltimore, MD, Aug 1995

Min BY, Wake B (2016) Parametric validation study for a hovering rotor using UT-GENCAS. AIAA-2016-0301. Presented at 54th AIAA aerospace sciences meeting, AIAA SciTech conferences, 4–8 Jan 2016, Sab Diego, California

Minot A, de Saint V, Marty J et al (2015) Advanced numerical setup for separation-induced transition on high-lift low pressure turbine flow using γ-$\text{R}\tilde{\text{e}}_{\theta t}$ Model. ASME, Montreal, 15 June 2015

Nichols R, Tramel R, Buning P (2006) Solver and turbulence model upgrades to OVERFLOW2 for unsteady and high-speed flow applications. AIAA Paper 2006-2824, Paper presented at the 25th Applied aerodynamics conference, 5–8 June 2006, San Francisco, California

Potsdam MA, Strawn RC (2002) CFD simulations of tiltrotor configurations in hover. In: Proceedings of the AHS 58th annual forum, Montreal, Canada, 11–13 June 2002

Potsdam MA, Yeo H, Johnson W (2004) Rotor airloads prediction using loose aerodynamic/structural coupling. In: Proceedings of the American helicopter society 60th annual forum, Baltimore, MD, 7–10 June 2004

Sheng C (2011) A preconditioned method for rotating flows at arbitrary Mach number. Model Simul Eng 2011(537464):17. doi:10.1155/2011/537464, ISSN: 1687-5591

Sheng C (2014) Predictions of JVX rotor performance in hover and airplane mode using high-fidelity unstructured grid CFD solver. In: Proceedings of the AHS 70th annual forum, Montreal, Canada, 20–22 May 2014

Sheng C, Allen C (2013) Efficient mesh deformation using radial basis functions on unstructured meshes. AIAA J 51(3):707–720. doi:10.2514/1.J052126

Sheng C, Zhao Q (2016) Assessment of transition models in predicting the skin frictions and flow field of a full-Scale tilt rotor in hover. In: Proceedings of the American helicopter society 72nd annual forum, 16–19 May 2016, West Palm Beach, Florida

Sheng C, Wang J, Zhao Q (2016a) Improved rotor hover predictions using advanced turbulence modeling. J Aircr. Online on 4 July 2016. doi:http://dx.doi.org/10.2514/1.C033512

Sheng C, Zhao Q, Hill M (2016b) Investigations of XV-15 rotor hover performance and flow field using U^2NCLE and HELIOS codes. AIAA-2016-0303, Paper presented at 54th AIAA aerospace sciences meeting, AIAA SciTech, 4–8 Jan 2016, San Diego, California

Spalart PR (2009) Detached-eddy simulation. Annu Rev Fluid Mech 2009(41):181–202

Spalart PR, Deck S, Sur ML et al (2006) A new version of detached-eddy simulation, resistant to ambiguous grid densities. Theoret Comput Fluid Dyn 20(3):181–195 Int J Heat Fluid Flow 30 (1):66–75

Wadcock AJ, Yamauchi GK (1998) Skin friction measurements on a full-scale tilt rotor in hover. In: Proceedings of the American helicopter society 54th annual forum, Washington, D.C., 20–22 May 1998

Wadcock AJ, Yamauchi GK, Driver DM (1999) Skin friction measurements on a hovering full-scale tilt rotor. J Am Helicopter Soc 44(4):312–319

Wissink AM, Sankaran V, Jayaraman B et al (2012) Capability enhancements in version 3 of the helios high-fidelity rotorcraft simulation code. AIAA-2012-0713, Paper presented at AIAA 50th aerospace sciences meeting, January 2012, Nashville, TN

Yoon S, Pulliam TH, Chaderjian NM (2014) Simulations of XV-15 rotor flows in hover using
    OVERFLOW. Presented at the fifth decennial AHS aeromechanics specialists' conference, San
    Francisco, CA, 22–24 Jan 2014

Zhao M, Xiao ZX, Fu S (2014) Predictions of transition on a hovering tilt-rotor. J Aircr 51
    (6):1904–1913